FOOD FIGHT

The Battle to Protect Our Food and Water Against Terrorism

by

Rebecca Hohlstein M.S.

Goblin Fern Press

Madison, Wisconsin

The author does not intend this book to be a substitute for advice from your own medical or health professional. *FOOD FIGHT, The Battle to Protect Our Food and Water Against Terrorism*, should not be used to diagnose or treat illness of any kind. If you or someone you know is experiencing disease symptoms of the illnesses described herein, seek the advice of a medical professional. The author and the publisher expressly disclaim responsibility for any adverse effects arising from the use or application of the information contained in this book.

Cover design: Ron Vyskocil, Thumbprint Graphics

Book design and typography:
Printing Services Management, Inc.

Indexing: Willard and Associates, Madison, WI

Illustrations by the author

Published by
Goblin Fern Press
A division of K. Henschel Enterprises, Inc., Madison, WI
www.goblinfernpress.com

Printed in the United States of America
First printing: November 2002

Although they have never really had one,
FOOD FIGHT
is dedicated
to my children,
Alexander and Tyler,
with many thanks for their patience,
support and understanding, and
their infinite tolerance of take-out meals.

Most of all, this, my first book,
is dedicated
to my husband Mark,
for all the reasons above and many more.

TABLE OF CONTENTS

PART ONE
Bacterial Threats to the Safety of Foods

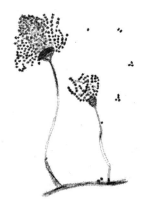

ACKNOWLEDGEMENTS

Heartfelt thanks are extended to: Dr. Robert Deibel, my graduate school advisor who instilled in me a love of the science of food bacteriology, launched my career, and mentored me along the way; Dr. Julie Zuehlke, dear friend and counselor who said, "You need to write a book!"; the Business and Professional Women of Madison for their encouragement; Anita Martin for spreading the word; Debra Morrill for connecting me and my publisher; Ron Vyskocil and Janet Pulvermacher for their wonderful cover designs, and Janet for turning a manuscript into a real book; and last but certainly not least, Kira Henschel, "Book Coach" and publisher, who took a chance on a new author, and guided me through the entire process.

INTRODUCTION

Perspective on Food Safety

When you sit down to a meal with your family, perhaps you take for granted that the food you are about to eat—the food your *children* are about to eat—is safe. We assume that the safety and wholesomeness of the foods we carefully choose, prepare and put on our tables is assured by those government agencies responsible for keeping an eye on things: the Food and Drug Administration (FDA), the United States Department of Agriculture (USDA), and the Food Safety and Inspection Service (FSIS). And in fact, these agencies do a good job: America's food supply is among the safest in the world.

Providing a safe food supply is hardly a new concept. Decades of research, monitoring, preventative government programs and regulations, and the development of various methods of food preservation have contributed to the safety of foods in the United States. Government agencies attempt to maintain this safety status by monitoring all aspects of food production, from farm to table, through surveillance, inspection and education.

But are these efforts enough? Even with current programs and safeguards in place, 76 million people annually fall victim to food-borne illness. Five thousand of these people die. And these statistics precede our newest threat: the deliberate contamination or destruction of our food supply. Following the tragic terrorist attacks of September 11, 2001 and the events that followed, our government's concern has extended to include safeguarding our food supply against such terrorist activity.

Prior to 9/11, the Centers for Disease Control and Prevention (CDC) issued a public health report in January of 2001, cautioning that the sabotage of our food and water is *the easiest means* of widespread terrorism. In February of 2002, the Food and Drug Administration (FDA) voiced similar concerns, stating that food could be used as a vehicle for a biological agent that would quickly reach huge numbers of Americans. Many countries including the United States, have developed germ warfare geared to attack the food supply.

Even before this new threat, infringement of government regulations by food manufacturers and processors has predictably and routinely occurred, either as a result of employee ignorance, indifference, or both. Government inspectors are spread very thin and are incapable of being in all plants at all times. Quality assurance personnel do their best to stay on top of things in their facilities, but like the government inspectors, they can only do so much.

Until recently, the worst food-based threats have been, with a few exceptions, limited to isolated outbreaks affecting only a few people and resulting in comparatively few deaths. These incidents are insignificant compared to the potential devastation from a deliberate terrorist sabotage of our food or water, a single act of which, according to the CDC, could potentially result in *hundreds of thousands of casualties*—astronomically more than those that resulted from September 11. With the millions of people employed in every facet of food production, sabotage by a single person would mean the end to our confidence in the safety of our food and water.

What are the chances of sabotage?

What can be done to thwart such efforts?

In a CBS News broadcast on November 11, 2001, Peter Chalk, a policy analyst with the RAND think tank stated, "Thousands of food processors nationwide lack proper security, and few test their finished products for contaminants." In the same news program, Robert Robertson of the U.S. General Accounting Office is quoted as saying, "We believe there is a reason to doubt our ability to detect and fully respond to an organized bioterrorism attack."

In January 2002, the FDA published *voluntary guidelines* for the producers, processors, transporters, and retailers of foods. These guidelines include criminal background checks and watchdog efforts for suspicious employees. There are also suggestions for improving security to prevent tampering. These recommendations are certainly a step in the right direction, but may be inadequate to deter the determined efforts of a terrorist group.

Secretary of Health and Human Services Tommy Thompson indicates that security of our food supply is his primary concern; it needs to be *everyone's* primary concern. Terrorism must not destroy our faith in the inherent safety of our food and water, which we *should* be able to take for granted. Where do we start to fight such an enemy? By educating ourselves about the potential agents of food bioterrorism and what can be done to prevent their use. Knowing our enemy will enable us to learn how to defeat it.

Ignorance is *not* bliss;
and
knowledge is power.

PART ONE

Possible
Bacterial Threats
to the
Safety of Foods

CHAPTER ONE

Bacillus anthracis: ANTHRAX

Anthrax, once primarily the scourge of farmers and veterinarians, has again reared its ugly head after decades of relative anonymity. This organism became the biological weapon of choice to continue the spread of terror that had been initiated with the events of September 11, 2001.

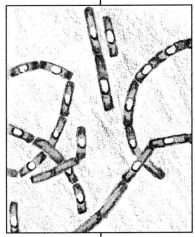

Bacillus anthracis with central spores

Anthrax is one of the most serious among the organisms identified by the *Working Group on Civilian Biodefense,* an organization made up of twenty-five representatives from academia, government, and public health, formed to review data on potential bioweapons and develop recommendations for the medical and public health professionals in the event that these weapons are ever used. The Working Group deemed anthrax capable of causing sufficient disease and death to cripple a city. Contamination of our country's mail was easily accomplished with a powdery concentrate of *B. anthracis* spores, resulting in several cases of anthrax. The same organism could potentially be used to contaminate food, the consumption of which would lead to an intestinal form of the disease, no less insidious and even more difficult to treat than the inhalational form.

The Organism

Bacillus anthracis is a very large rod-shaped bacterium, called a bacillus, whose species name comes from *anthrakis*, the Greek word for coal, due to the formation of black, coal-like lesions in the cutaneous (skin) form of the disease. *B. anthracis*, along with other species of *Bacillus*, is commonly found in soil, dust, water, decaying vegetation, and occasionally foods. *B. anthracis* has been found in soil samples from Asia, Africa, Central and South America, the Caribbean, the Middle East, parts of Europe, and in the United States, primarily along old cattle trails. Some other members of the genus *Bacillus* are also pathogenic (capable of causing illness), but most are harmless, and some even produce essential antibiotics. For example, *B. thuringiensis* is an insect pathogen that is used to protect crops.

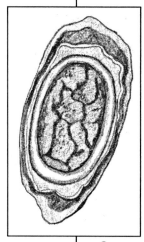

Cross-section of a bacterial endospore

Survival of *Bacillus* species in adverse environments is facilitated by their ability to form a protective structure called an endospore, or spore for short. (The only other bacterial genus capable of forming spores is *Clostridium*, which is the subject of Chapter Three.) In favorable environments, the vegetative form of the bacterium, which takes in nutrients and eliminates waste products, carries out these activities, as well as reproduction by cell division. But when nutrients become limited or the environment becomes inhospitable due to heat, cold, or lack of water, sporulation (spore formation) takes place. When this occurs, a spore forms within the bacterial cell, either in the center of the cell (central spore) or at the end (terminal spore).

Spores are the most resilient naturally-occurring structures known to science. They represent a suspended animation of sorts, a dormant state in which the bacterium can survive for extended periods of time—some spores have been recorded to be over fifty years old. Spores are resistant to heat, dehydration and other adverse conditions that would kill the vegetative form of the bacterium. Our modern sterilization procedures, which utilize a combination of heat and high pressure, were developed to destroy these tough structures. Spores are much more difficult to kill than the vegetative form of the bacterium, which can be destroyed with ordinary antimicrobial agents such as alcohol or or peroxide. In the case of anthrax, the spore is the infectious form.

Anthrax in History

Although believed to be the culprit of the fifth and sixth plagues of ancient Egypt, the earliest claim to fame for *B. anthracis* was due to the work of Dr. Robert Koch, a physician who studied disease transmission in cattle in the late 1800s. Dr. Koch isolated this organism from diseased animals and introduced it to healthy animals, which then also became ill. Subsequently he re-isolated the organism from the newly infected animals, thus proving that this bacterium was indeed the causative agent of anthrax. In 1876, he published this study in the form of *Koch's Postulates*, a set of criteria for proving that a specific microbe is the causative agent of a particular disease. His work contributed tremendously to the study of disease transmission, and to the microbiological techniques of isolation and cultivation of single bacterial strains still used today.

Between 1916 and 1918, the Germans used anthrax and glanders (discussed in Chapter Thirteen) to infect livestock and feed intended for export to enemy troops. But research involving use of anthrax as a biological weapon began over eighty years ago, and at least seventeen nations are now believed to have offensive biological weapons programs. It is not known how many of these nations are utilizing anthrax, but Iraq has admitted to producing and weaponizing this organism.

The potential use of anthrax as a biological weapon was tested by the British army during World War II. Spores were experimentally disseminated on an island off the coast of Scotland using an explosive shell as the means of dispersal. The spores persisted in the environment for thirty-six years. The area was effectively decontaminated in 1987.

Experiments performed by American scientists in the early 1950s involved dry runs made against cities evidencing similarities to potential Soviet targets. These included Saint Louis, Minneapolis and Winnipeg, which were the targets for the release of noninfectious aerosols to determine how much agent would be required to kill residents of Kiev, Leningrad and Moscow. Subsequent air samples showed that the agent traveled about a mile from its point of release.

Prior to September 11, 2001, the last time *Bacillus anthracis* made the news was in 1979, when an aerosolized form of the bacterium was accidentally released from a military microbiological facility in Sverdlovsk in the former Soviet Union. Seventy-nine people became ill with inhalational anthrax; sixty-eight of them died. This incident demonstrated that anthrax could indeed be used as a potent biological weapon. It also afforded epidemiologists the opportunity, albeit an unfortunate one, to study anthrax on a large scale.

Disease Manifestations of Anthrax

Typically, anthrax is a disease of herbivorous (plant-eating) animals. Domestic animals that most commonly contract anthrax include sheep, cattle, horses, and swine, although vaccination programs have greatly reduced its incidence. Humans get anthrax primarily through contact with an infected animal or animal product, including hair, hide or contaminated meat. Hence it has characteristically been associated with veterinarians and farmers.

Human anthrax is rare and manifests itself in three forms: cutaneous, inhalational, and intestinal. In all three forms, when death occurs, it is ultimately due to the organism spreading throughout the body and producing a deadly exotoxin, meaning that the bacterial cells secrete the toxin into their environment.

Up until 1954, death from anthrax was thought to be due simply to large numbers of bacteria in the blood that eventually blocked capillaries, restricting blood flow. This was known as the "log-jam" theory because examination of blood from animals dying of anthrax was found to contain 10^9 (1,000,000,000) bacteria per milliliter, high enough numbers to effectively dam up the minute capillary system.

However, it was later discovered that even lower numbers of cells in the blood (3 x 10^6 or 3,000,000/ml) could also result in death. Furthermore, cell-free plasma of anthrax-infected animals injected into normal guinea pigs initiated development of symptoms. These findings strongly suggested that *B. anthracis* produces a potent toxin that aids in its pathogenesis (an organism's innate ability to cause disease).

It is now known that full virulence (the ability to overcome the body's defenses) is due to the presence of a capsule, an outer covering that prevents the bacterium from being phagocytized (engulfed) by specialized cells of the immune system, and three toxin components. The capsule serves to protect the organism against this phagocytosis.

While the exact *modus operandi* of the toxins is not completely understood, death is the result of several factors, including oxygen depletion, and respiratory and cardiac failure. Once the toxin reaches a critical level, death is inevitable even if the bacterial cells themselves are eliminated with antibiotics.

Cutaneous Anthrax

From 1944 to 1994, 224 cases of cutaneous anthrax, the most common form of the disease, were reported in the United States. Disease usually follows exposure to infected animals and results in about two thousand cases annually. The largest known outbreak occurred in Zimbabwe, Africa, between 1979 and 1985 and resulted in over ten thousand cases.

Cutaneous infection results when a spore enters through a break in the skin. The first evidence of this type of infection is an itchy bump resembling a mosquito bite that becomes a liquid-filled sac (blister) after a day or two. Once inside this hospitable environment, the spore germinates (becomes vegetative once again) and produces the aforementioned toxin, which is responsible for local tissue death and the formation of the characteristic painless black scab. If not treated, cutaneous anthrax can become a systemic (body-wide) problem with a ten- to twenty-percent mortality rate. This form of anthrax, however, is easily diagnosed. While farmers and veterinarians are clearly at risk for contracting cutaneous anthrax, employees of slaughterhouses are also at risk if an infected animal is brought into the processing plant. The usual culprits of contamination of this type are swine, which may suffer chronic anthrax. If contamination occurs, elaborate measures must be taken to disinfect equipment, floors, and employees' clothing, arms and hands.

Inhalational Anthrax

Until recent terrorist activity, inhalational anthrax was primarily a concern for people employed by wool manufacturers, and has been dubbed "wool-sorter's disease." In the United States from 1900 to 1978, only eighteen cases of inhalational anthrax were reported, most of which occurred among wool sorters who contracted the disease from infected goat hair or skin, probably via aerosolized spores. In the 1960s, vaccination against anthrax became mandatory for those employed in goat hair mills.

As the name suggests, inhalational anthrax results when anthrax spores are inhaled and then settle into air spaces within the lungs. There, large white blood cells called macrophages (literally "large eaters") ingest the invading spores. Once inside the macrophages, some spores are subsequently destroyed. Surviving spores travel inside the circulating macrophages until they reach lymph nodes located in the chest, where germination takes place. After germination, the bacteria replicate and release the toxin that causes hemorrhage (bleeding), edema (swelling due to fluid accumulation) and necrosis (tissue death).

The stages of phagocytosis

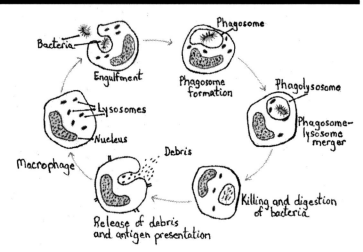

Initial symptoms of inhalational anthrax, which begin to show up two to sixty days following exposure, include headache, chills, chest pain, cough, fever, weakness and fatigue. These primary stages of the disease are easily confused with a wide variety of benign bacterial and viral infections that would not ordinarily warrant medical attention.

After a few hours to a few days, the secondary phase of infection ensues. This stage has a rapid onset characterized by enlarged lymph nodes, fever, labored breathing due to constriction of air passages, and profuse perspiration. Patients may also suffer meningitis accompanied with episodes of delirium and dulled reflexes. From this point, toxin accumulation and low blood pressure cause death within hours.

One of the most prominent concerns about inhalational anthrax is the potentially long incubation period, the time that elapses between inhaling the spores and exhibiting symptoms. Most cases will become symptomatic within a few days of when spore germination takes place, but spores may remain dormant within macrophages for up to eight weeks. This prolonged incubation adds to the difficulty of diagnosis. Cutaneous anthrax does not exhibit this latency period, most cases developing symptoms within twelve days.

Early diagnosis, though difficult, is also imperative because once symptoms start, death is most likely eminent within three days, even if antibiotics are started. This is due to the presence of the anthrax toxin, which is not affected by antibiotics.

Intestinal Anthrax

As is true with some other food-borne pathogens, anthrax of the ingested type results from the consumption of inadequately cooked, contaminated meat. While it is rare, outbreaks of this type have been reported in Asia and Africa, and two have been reported in Thailand: twenty-four cases in 1982 due to buffalo meat, and a second in 1987, of unknown source, resulting in fourteen cases.

Intestinal anthrax, similar to inhalational anthrax, results from the germination of ingested anthrax spores, but rather than germinating in the lymph nodes, they germinate in the upper or lower gastrointestinal tract. If germination takes place in the upper gastrointestinal tract, *oropharyngeal anthrax* (referring to the region of the back of the throat) results, with the formation of an ulcer, similar to that seen in a cutaneous infection, in the mouth or esophagus, accompanied by swelling of the neck. Symptoms may include fever and an inability to swallow.

In a lower gastrointestinal infection, *abdominal anthrax*, lesions form in the lowest part of the small intestine or in the beginning of the large intestine. Abdominal anthrax is accompanied by symptoms including loss of appetite, nausea, vomiting of blood, malaise, fever, abdominal pain and bloody diarrhea.

Advanced infection of the intestinal variety is similar to that of inhalational anthrax, resulting in enlargement of localized lymph nodes, edema, and finally sepsis, a toxic condition resulting from the spread of bacteria or toxin throughout the body. This can also occur in cutaneous anthrax, suggesting similar medical intervention. Intestinal anthrax, however, is more difficult to diagnose since its symptoms mimic those of food poisoning caused by an endless variety of other, less formidable bacteria. It is supposed, therefore, that even though the potential threat of intestinal anthrax resulting from terrorist sabotage is relatively small, death could result in twenty-five to sixty percent of cases.

Diagnosis and Treatment of Anthrax

Time is of the essence in the diagnosis of anthrax, but culturing (growing bacteria in the laboratory) can take up to several days. In this type of procedure, a sample of the patient's blood or sputum (saliva) is incubated in a medium (nutritious broth) conducive to bacterial growth. After a period of twenty-four to forty-eight hours, this enrichment is further analyzed, usually by plating on selective and/or differential media. Selective media inhibit the growth of unwanted microorganisms, while differential media have added ingredients that will change the media in some way, i.e., a color change, as the desired bacteria grow.

A standard blood culture is the most useful test to identify anthrax infection. Initial growth should be seen within twenty-four hours, but an additional one to two days are required for definitive diagnosis. Routine laboratory analysis often stops with identification of only the genus of the bacterium (genus *Bacillus*), but not the species (*anthracis*) unless directed otherwise. In the United States, most *Bacillus* species isolated in laboratories are *B. cereus*, a common food-poisoning organism. Given the current state of affairs, it is prudent to update lab protocol for early identification of *B. anthracis* as the causative agent of bacteremia (bacteria in the blood).

If *B. anthracis* spores are released in aerosol form, nasal swabs can be used to assay the extent of exposure among the population in a given area. Resulting information may be used to determine whether or not to administer prophylactic (preventative) antibiotics, but is not determinative of an infection in any one individual. In other words, a patient with a negative nose culture may still have anthrax in the lungs.

Anthrax rapid diagnostic tests, based on antigen-antibody interactions, yield results in hours but are not yet widely available. Currently, only national reference laboratories have these kits, which would be used to confirm a diagnosis, or to test the infecting organism's susceptibility to antibiotics. These tests would also be used to investigate anthrax hoaxes or to examine suspicious material found in the possession of a terrorist.

Medical X-ray examination of a patient with inhalational anthrax shows an enlargement of the mediastinum (the space in the chest cavity between the pleural sacs of the lungs), due to extreme swelling of lymph nodes. Fluid accumulation is also seen in the chest wall, but *B. anthracis* does not cause pneumonia (inflammation of the lungs). Fifty percent of advanced anthrax cases exhibit extreme hemorrhagic meningitis, an inflammation of the tissue surrounding the brain and spinal column.

The United States licensed an anthrax vaccine in 1970. This vaccine has been mandated for all active and reserve U.S. military but is in short supply. It will be years before enough can be made available for the general public. Even if enough were available, there are several reasons why countrywide immunization is not feasible, including the extreme cost accompanied by the unlikelihood of an attack in any given area. Also, it is not known how well immunization will work if the amount of anthrax spores inhaled or ingested is abnormally high. Nor is the vaccine suitable for children.

In the event of a terrorist attack with anthrax, early detection and administration of antibiotics would be the only recourse, without which the mortality rate could be as high as ninety percent. A delay in therapy of as little as a few hours could significantly lesson the chance of survival. Therefore, *if a known threat exists,* all exposed persons must begin taking antibiotics. Antibiotics should not be taken randomly as this will only proliferate the growing problem of antibiotic resistance seen with many bacterial pathogens.

Penicillin has been the antibiotic of choice for treatment of anthrax, but some penicillin-resistant strains do occur naturally and have been recovered from humans. Published reports indicate that one of the strains of *B. anthracis* used to produce a vaccine has also been engineered by Russian scientists to be resistant to penicillin and tetracycline, but the strain used to contaminate mail in September of 2001 did respond to treatment.

Anthrax strains developed specifically for use as weapons may have built-in resistance to conventional antibiotics. Therefore, alternate antibiotic therapy is recommended even though unsubstantiated reports of strains resistant to other antibiotics also exist.

Some types of *B. anthracis* also show a natural resistance to many antibiotics, so these should not be used for treatment or prophylaxis.

It is also important to note that studies have shown that when infected animals were given antibiotics for treatment of anthrax, they failed to develop their own immune response. For this reason and the potential for latent germination and disease development, therapy should be continued for a minimum of sixty days.

Person-to-person transmission does not occur with anthrax, so patient isolation and immunization of patient contacts are not necessary. However, animals and humans who have died of anthrax should be cremated. Special risks are involved with embalming and autopsies.

Studies of antibiotic treatment have been done with regard to inhalational anthrax, but the same types of antibiotics would suffice for ingested anthrax as well. The problem of early detection, however, remains for both types of infection. Development of vaccines and antibiotic therapy is ongoing.

Anthrax as a Biological Weapon

From 1989 to 1995, the number of nations working to develop biological weapons increased from ten to seventeen. Exactly how many of these are working with anthrax is not known. Due to the fact that anthrax is not contagious, it has been referred to as the perfect weapon of bioterror since only those directly exposed will become ill. Their inherent resilience adds to the effective use of *B. anthracis* spores as a biological weapon. These spores are stable in the environment for a protracted time period, so secondary aerosolization is also possible. Secondary aerosolization could occur if spores that have settled following primary distribution are somehow disturbed and again become airborne.

A lethal dose of anthrax is considered to be 10,000 spores. A single gram of dried powder could contain 10,000,000 spores or more. Less than one millionth of a gram is invariably fatal. According to the Office of Technology Assessment of the U.S. Congress, given the proper environmental conditions—clear, calm skies—220 pounds of aerosolized anthrax released over Washington D.C. could kill up to three million people, lethality equal to that of a hydrogen bomb. Similarly, if 110 pounds of anthrax powder were released from an aircraft along a 1.25-mile line, it would travel to contaminate about 12.5 miles. This lethal cloud of anthrax would be invisible, odorless and colorless, and would go undetected by atmospheric warning systems. Because the spores are so small, being indoors would offer no additional protection. The Centers for Disease Control and Prevention estimated that such an attack would cost the United States $26.2 billion per 100,000 people exposed.

Extensive knowledge and equipment are necessary to produce these lethal powders. *Bacillus anthracis* is first grown in the laboratory and then induced to form spores. The resulting culture is then concentrated and dried into a fine, easily dispersed, easily inhaled, and relatively stable powder.

One of the methods of producing this dry powder is called lyophilization (freeze-drying). In this process, water is removed from the culture by direct transfer from the solid state (ice) to the gaseous state (water vapor) through the use of temperature and pressure gradients that allow this transition of water molecules without passing through the liquid phase. Many types of bacterial cultures, as well as various food products, are preserved by this method.

Spray drying is another possible means of preservation as a dry powder. It involves subjecting a spray of the concentrated liquid culture to a stream of hot, moving air. This process is more detrimental, however, resulting in greater cell damage than freeze-drying.

Virtually any bacterial species can be prepared using these methods. Many types of bacteria are processed in this manner for legitimate use, including silage inoculants, starters for cheese manufacture, and probiotics (bacteria taken as health supplements) such as *Lactobacillus acidophilus*, which are used for maintenance of intestinal health and other purposes.

As previously stated, the anthrax strain used to contaminate the U.S. mail was sensitive to antibiotics, whereas most bacteria developed as bioweapons would probably be antibiotic resistant. Some evidence did exist, however, that this anthrax had been weaponized to increase human susceptibility to infection. This was accomplished by treating the spores to make them more dispersible in the air, and thereby easier to inhale. Other samples had an added irritant to accommodate the attacking of skin. Additional characteristics of anthrax as a weapons-grade biological agent include its small size (1-3 microns), lack of clumping, the number of spores per gram, and an effective delivery system.

The source of the anthrax used to contaminate the mail after September 11, 2001, remains unknown and is still under investigation by the Federal Bureau of Investigation. Authorities believe that whoever developed this weapon was well versed in bacteriological methods in order to have produced the weapons-grade anthrax, considered to be the best ever made. The culprit could be foreign or a U.S. citizen. However, the FBI is tending toward a domestic terrorist, possibly a disgruntled government employee with an expertise in anthrax, but less than fifty people in the U.S. fit this profile.

Recent indications that the anthrax may have been produced abroad and brought into the United States by al-Qaida terrorists include the discovery of a lab near Kandahar, Afghanistan, with some of the necessary equipment and trace amounts of anthrax. Two of the terrorist hijackers received treatment for cutaneous anthrax. Richard Spertzel, a former head of the U.N. Special Commission Biological Weapons Inspections Force, indicated that the quality of anthrax used in the attacks was too good to be anything but a foreign source, a nation with an advanced bioweapons program.

Anthrax as a Contaminant of Food or Water

A review of the events that took place in Sverdlovsk, in the former Soviet Union, in 1979 leads one to believe that inhalational anthrax would be responsible for the most cases and the most deaths in the event of a terrorist attack with spores. Little information exists regarding direct contamination of food and water with *Bacillus anthracis* spores. Since inhalational anthrax is the most deadly form of the disease, terrorists would most likely use this weapon as an aerosol. However, direct contamination of water or food is well within the realm of possibility.

Encouraging research has shown that deliberate intestinal infection of primates has failed to cause illness, even though naturally acquired ingested anthrax has been known to cause human fatalities. As with all types of pathogenic microbes, whether or not illness ensues depends on the general health of the victim and the size of the infecting dose.

As previously stated, anthrax is not contagious, and it is not likely that it would be carried into your home on clothing or skin. Any person coming in direct contact with alleged anthrax should thoroughly wash exposed skin and clothing. As for contaminated food or water, be aware of the symptoms of intestinal anthrax, but do not use antibiotics prophylactically. This can lead to serious complications, including side effects and antibiotic resistance in more common infectious bacteria.

The medical community is continually acquiring new knowledge of *B. anthracis* and its pathogenesis, diagnostic techniques and treatments, including a vaccine. An accelerated vaccine development effort is needed for an improved second-generation product requiring fewer doses. Maximum incubation times also need to be determined for the inhalation and ingested forms of anthrax in order to establish the necessary duration of post-exposure antibiotic therapy.

Bacillus anthracis

Disease	Symptoms	Transmission	Prevention / Treatment	Food / Water Sabotage
Cutaneous anthrax	Blister or black scab on skin	Open cut or skin abrasion	Avoid infected animals, wash hands / Antibiotics	No / No
Inhalational anthrax	Flu-like followed by respiratory distress	Inhale spores	Avoid inhaling spores, wash hands / Antibiotics	No / No
Intestinal anthrax	Ulcer in mouth or throat, difficulty swallowing, vomiting, diarrhea	Ingest contaminated/ undercooked meat	Thorough cooking of meat / Antibiotics	Possible / Possible

CHAPTER TWO

Salmonella: Food Poisoning

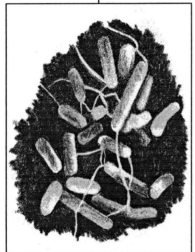

*Salmonella
enteritidis*

In September and October of 1984, in the small Oregon city of The Dalles, over 750 people became ill with gastroenteritis after eating in local restaurants. Investigations led to the discovery that salad bars had been purposely contaminated with the bacterium *Salmonella typhimurium* by members of a religious commune. This outbreak gave *S. typhimurium* a proven track record as an effective bioterrorist's weapon. Its victims virtually overwhelmed the city's health care facilities.

Salmonella has been recognized as a food-borne pathogen for over one hundred years, but new sources continually turn up, including unpasteurized apple cider and orange juice, alfalfa sprouts, cantaloupe, tomatoes, and even toasted oat cereal. The United States reports 40,000 cases of salmonellosis each year, most occurring among children. Mild cases are usually not reported or even diagnosed, so the actual number may be as high as two to four million annually. The CDC estimates that one thousand people die of *Salmonella* infections each year.

The Organism

Members of the genus *Salmonella* are rod-shaped bacteria and were therefore originally placed in the genus *Bacillus*. They are very widespread in nature and occur in the intestinal tract of virtually all types of animals, including insects. Reptiles and amphibians are notorious for carrying *Salmonella* and should not be kept as pets in preschools or daycare centers, as young children are very susceptible to *Salmonella* infections. Everyone should thoroughly wash their hands after handling reptiles or amphibians. In 1975, the sale of small turtles was halted in the U.S. because they often harbored *Salmonella*.

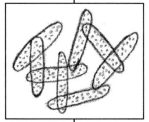

These bacteria are also common water and soil organisms, though they are not recognized as being free-living (capable of surviving outside a host for any length of time.) Most are motile (able to move) because of the presence of whip-like structures called flagella. Antigenic structures, which trigger the formation of antibodies by the human immune system, exist on various parts of the bacterium. Antigens on the flagella ("H" antigens) and the cell wall ("O" antigens) serve as the means for classification of the members of this genus into serotypes, or classifications based on antigen-antibody reactions. The species *Salmonella enterica* is divided into six subspecies collectively containing over two thousand serotypes. Naming of recently characterized serotypes has been standardized to reflect the location of their discovery, i.e. *Salmonella london* and *S. miami*.

Agglutination of "O" antigens

Salmonella are facultative which means they can live with or without oxygen. This, along with their tolerance to environmental extremes, makes them adaptable to a wide variety of environments. *Salmonella* are able to *grow* in temperatures ranging from 48°F to 117°F, salt concentrations up to 6.5 percent, and within a pH (acidity) range of 4.0 to 9.0 (on a scale of 1 to 14). They are able to *survive* conditions beyond these, including very acidic environments and freezing. Viable *Salmonella* have been isolated from apple juice at pH 2.0 and from highly salted beef jerky. These organisms are destroyed by thorough cooking. *Salmonella's* unique characteristics have made it possible for it to become one of the most prevalent causes of food-borne illness,

Agglutination of "H" antigens

secondary only to *Campylobacter jejuni,* a diarrhea-causing bacterium commonly found in raw chicken and raw milk. All *Salmonella* are considered to be potential pathogens, and therefore subject to "zero tolerance" levels in processed foods, meaning that if routine quality assurance testing finds even a single *Salmonella* bacterium, the food cannot be released from the processing plant.

Historical Perspectives of Salmonella

The most famous *Salmonella* story is that of Typhoid Mary. Mary Mallone lived in Long Island, New York, in the early 1900s. She was a carrier of *Salmonella typhi,* the bacterium that causes typhoid fever. (A carrier is an individual who harbors the organism in the digestive tract, but does not exhibit symptoms of illness.) As luck would have it, Ms. Mallone's line of work involved preparation of food for private homes in the Long Island area. Through her employment and lack of good personal hygiene, she managed to infect twenty-six individuals from seven different families. She supposedly retired in 1910, but was once again discovered preparing food in 1912, at which time she was arrested and hospitalized.

A half century earlier, typhoid fever of epidemic proportions plagued both Union and Confederate soldiers during the Civil War. When the war ended, typhoid fever survivors vectored the illness to all parts of the country, greatly increasing its incidence until sanitation and proper hygiene methods were developed to prevent further spread of the disease.

Salmonella was first identified as an agent of food poisoning in 1888, when it was isolated by a German scientist, August Gartner, from an individual who had died after eating beef from an infected cow. The same organism was isolated from the remains of the cow. After this initial association, the organism became commonly known as an agent of food-borne illness. It was eventually renamed after American veterinarian D. E. Salmon who, in collaboration with Theobold Smith, isolated a similar bacterium, *Bacillus cholera suis,* a secondary invader in the viral disease known as hog cholera. This group of bacteria became alternatively known as the Gartner group, the hog cholera group, or the salmonella group.

The most noteworthy outbreaks of food-borne salmonellosis are quite recent. In the spring of 1985, a Chicago dairy plant processing 1.5 million pounds of milk a day suffered irreparable damage when a faulty pipeline enabled raw milk to seep into lines carrying pasteurized milk, thereby contaminating the pasteurized milk with *S. typhimurium*. This incident caused 16,285 confirmed cases of salmonellosis, at least two deaths, and the closing of the plant.

In 1994, 224,000 people throughout the Midwest contracted salmonellosis after eating ice cream contaminated with *S. enteritidis*. Tanker trucks that had previously been hauling unpasteurized, liquid eggs were subsequently used to transport ice cream premix to the manufacturing plant. The trucks were cleaned and sanitized between loads, but outlet valves were not thoroughly disinfected. The premix was not pasteurized again before entering the production line, a process that would have killed the salmonellae organisms.

Salmonella Infections: Disease Manifestations

Salmonella infections can be divided into two basic types: enteric fevers originating from the intestine, the most serious of which is typhoid fever, and gastroenteritis, an inflammation of the gastrointestinal tract. Humans are the only reservoir for the salmonellae that cause enteric fevers, which are spread via food or water that has been contaminated with human waste. Therefore, enteric fevers are more prevalent in underdeveloped countries lacking proper waste disposal systems, but they do occur rarely in developed countries as well. Enteric fevers can also be spread directly by food handlers not practicing good personal hygiene, as was the case with Typhoid Mary.

The serious nature of typhoid fever is due to the ability of *S. typhi* to penetrate the intestinal wall and enter the bloodstream. Such an organism is said to be *invasive*. As few as fifteen to twenty ingested cells are enough to start an infection, which begins when bacteria attach to intestinal epithelial cells, break down cell walls and enter the cells themselves. The invasive action deteriorates the lining of the intestine, resulting in bloody diarrhea. One to two weeks later, the organism makes its way into the bloodstream. The range of symptoms expands to include headache, high fever, and vomiting. Once the bacteria are in the bloodstream, the infection quickly becomes systemic and can affect virtually any organ of the body. The fatality rate of typhoid fever is ten percent. A less severe form of enteric fever, known as paratyphoid fever, is caused by *S. typhimurium* or *S. hirschfeldii*.

The most common manifestation of *Salmonella* infection is the gastrointestinal form, which is caused by consumption of foods contaminated with *Salmonella* serotypes other than those that cause enteric fever. The majority of outbreaks are due to the mishandling of foods in the home or food service establishments. Because of its presence in animal intestinal tracts, *Salmonella* is a common contaminant of beef, turkey, chicken, pork, milk and milk products, peanut butter, chocolate, and raw eggs. Foods coming into contact with contaminated water can also harbor *Salmonella*, including vegetables, fruits, rice and seafood. Virtually any type of food can be contaminated directly by an infected food handler. Although 10,000,000 to 1,000,000,000 organisms per gram are usually required for illness, the number may be as low as three organisms per gram of food.

Acute gastroenteritis usually lasts one to two days, but can persist longer depending on how many bacteria are consumed, the host's immune response, and the potency of the infecting *Salmonella* strain. Symptoms are diarrhea, cramps, nausea, vomiting, fever and headache. These ensue anywhere from five to seventy-two hours after consumption of contaminated food or water, and may last up to a week. Individuals whose immune systems are compromised in some way—the very young, the very old, AIDS or transplant patients—are more likely to suffer an invasive form of the disease and may require medical intervention. Although the death rate is only one percent for the general population, most deaths due to *Salmonella* are among the elderly in nursing homes, where the rate increases to 3.6 percent.

A few people (two percent) having recovered from a *Salmonella* infection will go on to suffer arthritis, eye irritation and painful urination, a condition known as *Reiter's syndrome*, which can last months or years and may lead to chronic arthritis. Other long-term effects include aneurisms (dangerous thinning of the walls of veins or arteries), infections of the heart, spleen or liver, and meningitis.

Salmonella Contamination of Eggs

S. enteritidis is the most common cause of salmonellae infection in the U.S. and has been problematic in raw eggs since the 1980s. The problem began in the Northeast, but by 1989, sixteen states had reported egg-related outbreaks of *S. enteritidis*. It was once believed that this organism contaminated only the surface of the egg, via fecal material from the hen, at the time the egg was laid. Stringent procedures for the cleaning and inspection of eggs, implemented in the 1970s, have made this type of contamination rare. The current epidemic is due to *S. enteritidis'* presence inside whole, Grade A, disinfected eggs.

Contamination of the interior of the egg was previously thought to occur when surface organisms penetrated the shell. It has since been established that contamination is by "transovarian" or "vertical" transmission, wherein *S. enteritidis* actually enters the egg as it is developing within the ovary of an infected hen. The inside of the egg will still appear normal. Only a small number of hens are infected at any given time, and an infected hen will only occasionally produce a contaminated egg, while the majority will be salmonella-free.

In the Northeast, it is estimated that one egg in 10,000 may be internally contaminated. Nationwide, this decreases to one in 20,000, meaning that one in fifty people in the U.S. will be exposed to an *S. enteritidis*-containing egg each year, but the mere presence of a *Salmonella* bacterium in a single egg does not guarantee illness. Since the organism is killed by heat, if the egg is thoroughly cooked (not sunny-side up!), the bacteria will be destroyed. But when eggs are pooled into large batches for use in restaurants or other food service establishments, the contaminant of one egg becomes the contaminant of five hundred. If the eggs are not handled and stored properly, the organism will be able to multiply.

From 1985 to 1998, there were almost eight hundred outbreaks of *S. enteritidis* reported to the CDC. These outbreaks were responsible for over 28,000 illnesses and seventy-nine deaths, and approximately eighty-two percent of them were due to eggs. In 1998, President Clinton established the Council on Food Safety, which developed the Egg Safety Action Plan to reduce and eventually eliminate eggs as a source of *S. enteritidis*. The plan calls for a fifty-percent reduction in the incidence of *S. enteritidis* in eggs by 2005, and complete eradication by 2010.

Some of the measures being taken include testing of breeder flocks that produce egg-laying hens to be sure they are *Salmonella* free. The United States Department of Agriculture (USDA) has issued guidelines for the proper handling and storage of eggs in retail food establishments, and also monitors *S. enteritidis* infection in egg-laying flocks.

There are *S. enteritidis* vaccines available that can be administered to baby chickens via drinking water. The vaccine confers lifelong immunity to the chickens and also protects eggs. This vaccine may eventually be appropriate for other animals and humans as well.

Diagnosis and Treatment of *Salmonellosis*

Blood or stool cultures are used in diagnosing *Salmonella* infections. Conventional culturing methods take five days just for presumptive results, but several rapid methods are available. A polymerase chain reaction (PCR) assay, a testing procedure that amplifies small fragments of DNA so they can be characterized, has recently been developed that will detect *Salmonella* in foods. Isolates are serotyped for epidemiological purposes.

Most cases of noninvasive salmonellosis will resolve themselves in five to seven days without antibiotic therapy. Antibiotic treatment will not shorten the duration of illness, and has no bearing on whether or not the patient will go on to develop arthritis or other long-term complications. Antibiotics *should* be administered to patients who are at high risk for developing, or have already developed, the invasive form of the disease. Twenty to thirty percent of *Salmonella* isolated from infections are resistant to antibiotics, largely due to the use of antibiotics to promote growth of animals used for food. Many *S. typhimurium* isolates from the U.S., Canada and Europe have been found to be resistant to several antibiotics. Additionally, *S. typhi* bacteria can hide inside host macrophages, where they are somewhat protected against antibiotic action, so ongoing therapy may be necessary.

Preventing *Salmonella* Contamination in Foods

Investigations indicate that the majority of *Salmonella* outbreaks are due to eggs or egg-containing dishes that are either raw, undercooked, or held at room temperature.

Raw poultry often carries this organism as well, and foods contaminated with *Salmonella* will look and smell normal. These bacteria are destroyed by heat, so avoid raw or undercooked eggs, poultry or meat. Beware of dishes that may contain raw eggs, such as homemade ice cream or mayonnaise, Hollandaise sauce, Caesar dressing, or cookie dough. Do not drink juices that have not been pasteurized. If *Salmonella* bacteria are present in heat-processed foods, it is due to either post-processing contamination, or insufficient time/temperature conditions in the heating process itself.

Many cases of *Salmonella,* as will as other types of food poisoning, are due to cross contamination. Raw meats and raw eggs must be kept separate from other foods that are not going to receive heat treatment, such as vegetables for a salad. Cooked foods must be kept hot and served promptly, or cooled quickly for refrigerated storage.

Since *Salmonella* is present in the digestive tracts of animals, natural manure fertilizers can serve as a source of contamination for fruits and vegetables, which should be washed thoroughly if they are to be eaten raw.

Public health departments must be notified of cases of *Salmonella* food poisoning, and isolates should be serotyped to trace the source of infection. Government-initiated improvements in practices of farm animal hygiene and slaughter, and fruit and vegetable harvesting and packaging, are aiding in the prevention of contamination by *Salmonella* and other pathogens. Better education of food industry and food service personnel, and the use of pasteurized liquid eggs in restaurants, hospitals, and nursing homes will also reduce the incidence of *Salmonella* infections.

Salmonella as a Biological Weapon

After President Nixon's 1969 ban on research into biological weapons for offensive use, it was discovered that the Central Intelligence Agency had kept enough pathogens and toxins to effectively kill or sicken millions of people. Two types of *Salmonella* were among the stash.

This organism has already been proven to be a suitable weapon of food sabotage, as was demonstrated in Oregon in 1984. That outbreak also exemplifies the ease with which self-service foods can be contaminated with *Salmonella* or other pathogens.

The religious cult responsible for the Oregon outbreak, the Rajneeshee, also put *S. typhimurium* into cups of drinking water, and attempted to poison public water systems. They sprinkled the bacteria on lettuce in grocery store produce departments, and considered injecting it into milk cartons, but feared there would be overt evidence of tampering. *Salmonella* was, however, used to taint coffee creamers, salad dressings and salad bars in restaurants.

Criminal investigations a year after the outbreak discovered that the cult had ordered a multitude of other pathogenic bacteria from the American Type Culture Collection, a bacterial-supply house located in Virginia, including *S. typhi* and *paratyphi, Francisella tularensis, Enterobacter cloacae, Neisseria gonorrhoeae,* and *Shigella dysenteriae.* If *S. typhi* had been their weapon of choice, there would have been at least a few deaths among the 751 victims.

It is clear from the Oregon scenario that it will be difficult to protect the public against deliberate contamination of self-service foods, even today. As was previously stated, the vaccine currently used to protect egg-laying chickens against *S. enteritidis* may someday be available for humans, but is not currently on FDA's list for vaccine development. There is a vaccine for prevention of typhoid fever, however. It is recommended for those traveling to areas of high-risk for fecally contaminated food or water, including countries in Africa, Asia and Central and South America, or for individuals coming into continual contact with a known carrier.

The best defense is to be familiar with symptoms of *Salmonella* poisoning, and to seek prompt medical attention if necessary. Unless a strain of "super-*Salmonella*" is developed by terrorists, those individuals most vulnerable to infection will continue to be children younger than eight years, the elderly, and the immunologically compromised.

Though *Salmonella* is a dangerous pathogen that warrants our full attention in food processing, food service, and our own kitchens, it is not among those organisms deemed most likely to be used as weapons of terror.

Salmonella				
Disease	**Symptoms**	**Transmission**	**Prevention / Treatment**	**Food / Water Sabotage**
Typhoid fever (*S. typhi*)	Bloody diarrhea, fever, vomiting	Food or water contaminated with human fecal material	Avoid drinking water in areas with poor sanitation; vaccine / Replace fluids	Possible / Not probable
Paratyphoid fever (*S. typhimurium*)	Same as above but less severe	Same as above	Same as above	Same as above
Salmonellosis (*S. enteritidis* & others)	Diarrhea, vomiting, fever	Foods, especially meat & eggs	Thorough cooking; prevent cross-contamination / None or antibiotics	Possible / Not probable

CHAPTER THREE

Clostridium: Botulism & Tetanus

The bacterial species discussed in this chapter, though members of the same genus, vastly differ in the severity of the diseases they cause. Yet due to the toxins they produce, all possess potential for use as biological weapons, and all are easy to come by—they exist right in your own back yard.

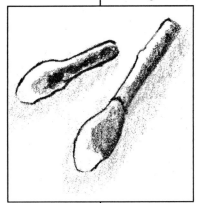

Clostridium tetani

Botulism strikes terror in the hearts of scientists and lay people alike. Cases of food poisoning by *Clostridium botulinum* receive worthy notoriety since they are often fatal. This organism is capable of producing one of the most toxic substances known, a neurotoxin ten times more deadly than rattlesnake venom; one-trillionth of a gram of pure botulinum toxin can kill a healthy adult. Therefore, it has long been recognized as a potential biological weapon.

Its cousin, *Clostridium perfringens,* provides a little versatility in the diseases it incurs, ranging from gas gangrene to gastroenteritis. It is also among the most common food-poisoning culprits, causing an estimated 10,000 cases annually in the U.S. *C. perfringens* causes sickness by the production of toxins as does *C. botulinum*, though of a different nature. These most potent of toxins have serious implications as weapons of terror.

The Organism: Genus *Clostridium*

Member of the genus *Clostridium* are rod shaped and have the unique characteristic of being strict anaerobes, unable to grow in the presence of oxygen. Clostridia are very common and distributed throughout the world, particularly in anaerobic pockets in soil where they live as saprophytes which fix nitrogen and decompose cellulose. They are also normal residents in the intestinal tracts of a wide variety of animals, including humans. Dense populations of *C. perfringens* occur where animals congregate, such as grazing and watering areas.

Pathogenic species infect humans usually via contaminated vegetables or meats. *Clostridium* is the only other bacterial genus besides *Bacillus* to forms spores, the means by which foods are contaminated. Like all spores, *Clostridium* spores are very resilient—those of *C. perfringens* can survive five hours in boiling water. Since spores are prevalent in soil, they readily contaminate plant products and animal hides, and subsequently foods.

C. tetani is another member of this genus. This organism is the causative agent of tetanus, commonly associated with stepping on a rusty nail. Tetanus is actually caused when the bacterium, which is present in the soil, gains access to the body via a deep puncture wound. A painful, once-per-decade tetanus shot keeps the organism from growing in wounds deep enough to exclude oxygen. If infection does ensue, it remains localized, but the toxin produced by *C. tetani* spreads, causing systemic neurological paralysis. Toxin adheres to nerve synapses, which in turn causes simultaneous contraction of opposing muscles. "Lockjaw" results when muscles of the mouth are affected. Involvement of muscles of the diaphragm causes death by suffocation.

Clostridium in History

From 886 to 912 A.D., Emperor Leo VI of Byzantium forbade consumption of blood sausage (pig intestine filled with blood and other ingredients) due to its harmful health effects. The organism responsible for this order was most likely *Clostridium botulinum*. In 1793, thirteen cases and six deaths in Germany were traced to the same type of sausage. Hence the organism got its name from the Latin word for sausage, *botulinus.*

Three deaths in 1895 in Belgium led to an investigation and discovery of *C. botulinum* in raw salted ham, after twenty-three musicians attending a funeral became ill. The largest recorded out-break of botulism occurred in 1933 in Russia due to contaminated stuffed eggplant; 230 became ill and ninety-four died. Because of modern food processing parameters, the incidence and size of bot-ulism outbreaks have greatly decreased.

Clostridium perfringens has been known since 1898, but was first recognized as a contaminant of foods in 1943 after it was discov-ered that anaerobic organisms could grow in foods. By the 1950s, it became a frequently reported cause of food-borne illness in the United States and Europe, and is now recognized as one of the leading causes of food poisoning.

It is also the causative agent of ninety percent of the cases of gas gangrene, a wound infection characterized by muscle tissue death accompanied by odiferous gas production. A wound accompanied by tissue damage so severe that circulation is interrupted creates the anaerobic conditions necessary for growth and toxin produc-tion. *C. perfringens* invades the wound and liquifies muscle tissue. In one to three days, infection spreads into subcutaneous tissue, causing necrosis (tissue death), fever, toxemia (toxin in the blood), shock, and death. Even with antibiotic therapy, amputation may be needed to halt spread of infection.

Disease Manifestations of Clostridial Food Poisoning

Clostridium perfringens

Because of their wide distribution in nature, clostridia are frequent contaminants of many types of foods including meat, poultry, fish, and vegetables. Spores of these organisms are present in a large percentage of meat products. Contamination occurs at the time of slaughter when spores are transferred from hides to meat surfaces. *C. perfringens* will be present in low numbers in raw foods of all kinds. Such low numbers can be ingested without causing illness, and numbers in the large intestine of healthy humans are about ten thousand per gram of feces.

Food poisoning by *C. perfringens* is usually associated with large quantities of food prepared in advance and held for later con-sumption, so outbreaks are most frequently associated with institutions, restaurants, and cafeterias. Commonly implicated foods include beef, chicken, turkey, gravies, stews, and creamed dishes made with meats.

In order for food poisoning to occur, conditions must be appropriate for the small numbers of spores normally present in the foods to germinate and proliferate. Spores are induced into germination by the initial cooking of the foods, which also depletes oxygen, thereby making a hospitable environment for the growth of these anaerobic organisms. If the prepared foods are not held at proper temperatures, i.e., not properly refrigerated or held at insufficiently hot temperatures, the bacteria will multiply.

Foods mishandled in this way are said to be "temperature-abused" and may exhibit bubbling due to gas formation by rapidly multiplying bacteria. (If the baked beans are bubbling but are not hot, don't eat them!) Heating foods to 145°F will destroy the vegetative cells and the toxins.

C. perfringens causes one of the most serious types of food poisoning, *toxicoinfection,* a combination of an infection and toxin production. To cause *C. perfringens* poisoning, one million to ten million live bacterial cells must be ingested. These cells take up residence in the small intestine and continue to reproduce until they sporulate, at which time enterotoxins (toxin affecting the digestive system) are produced. Eight to twenty-four hours after the food is eaten, enterotoxins are released as vegetative cells lyse (break open), causing the symptoms of *C. perfringens* poisoning—intense stomachache, flatulence and diarrhea—which usually last about twenty-four hours. This type of food poisoning is often mistaken for the twenty-four-hour flu.

The enterotoxin acts on the small intestine, binding to the brush border (outermost) epithelial cells, changing their permeability and disrupting osmotic balance. An excess of sodium and chloride ions causes fluid to accumulate in the lumen of the intestine, which leads to diarrhea. As previously stated, the human intestine normally harbors around ten thousand *C. perfringens* per gram of fecal material, but during an episode of toxicoinfection, there will be one million or more cells per gram of feces.

Severe diarrhea can lead to dehydration. Fatalities due to such dehydration are very rare and usually limited to elderly or debilitated persons. Sometimes *C. perfringens* poisoning is very mild, and often goes unreported so the true incidence is unknown. Only outbreaks involving several people are reported.

A more serious illness, necrotic enteritis, is caused by a different strain of *C. perfringens.* This disease is rare, but continues to be a problem in developing countries from consumption of improperly prepared pork. It occurs occasionally in developed countries, where it is associated with chitterlings (pig intestines). As *C. perfringens* grows in vivo (in the body), it produces a toxin that destroys intestinal tissue.

Clostridium botulinum

Naturally occurring botulism is serious but rare, with fewer than 200 cases reported annually in the United States. Manifested in three forms, *infant, wound,* and *intestinal,* botulism is caused by an exotoxin (one that is secreted into the bacterium's environment). One milligram is enough to kill over one million guinea pigs.

There are seven different strains of *C. botulinum,* designated A thru G, each producing a characteristic toxin type. Types A, B, E and F can cause illness in humans, type A being the most common, F the least. Toxin types A and B are associated more so with vegetables and fruits, while E is associated with aquatic environments, especially marine, where it proliferates in dead animals and sediments and is disseminated by water currents and migrating fish. Types C and D cause illness in birds and foraging animals, respectively, but not in humans. However, it is thought that humans may be susceptible to C and D because these strains are able to cause botulism in primates. Type G is not known to cause food-borne botulism.

All botulinum toxins are neurotoxins that bind to specific sites of nerve synapses and prevent the release of acetylcholine, the chemical that signals muscle contraction. As a result, muscles remain constantly relaxed, a condition known as flaccid paralysis. Some individuals may be only slightly affected, while others will require months of care. Severity is dependent upon the amount of toxin ingested and absorbed into circulation. There is no impairment of mental function, but patients may have difficulty communicating due to facial muscle paralysis.

Flaccid paralysis is progressive. Six hours to two weeks (most commonly twelve to thirty-six hours) after ingesting toxin-containing foods, symptoms ensue, starting with the muscles of the face. Paralysis creates blurred and double vision, difficult swallowing and droopy eyelids. These initial symptoms can also be accompanied by nausea and vomiting. As the paralysis descends, muscles of the shoulders, upper and lower arms are affected, followed by thigh and calf muscles as are heart and respiratory muscles. Undiagnosed botulism causes death by asphyxiation and/or heart failure. Prior to 1950, reported cases had a mortality of sixty percent. With modern care, this has been reduced to less than five percent.

Infant botulism is the most common type. It occurs in children less than one year of age when spores of *C. botulinum* are ingested and subsequently produce toxin in the intestinal tract. The source of spores can be something the baby put in its mouth that is contaminated with *C. botulinum*-tainted dust or soil, but it is usually syrup or honey used as a "pacifier dip." In adults, well-established intestinal flora successfully competes with *C. botulinum* to prevent its growth, but in young children, colonization of these beneficial bacteria has not yet taken place. As it grows in the intestine, *C. botulinum* produces symptoms that include constipation, weakness, poor feeding, and loss of head control, so infant botulism is often referred to as "floppy baby syndrome."

Wound botulism is the rarest form. In this disease, *C. botulinum* grows and produces toxin in a deep wound, evidencing symptoms similar to food-borne botulism.

The presence of the toxin in foods, rather than the bacteria themselves, leads to intestinal botulism. Similar to *C. perfringens*, *C. botulinum* spores contaminate raw foods prior to harvest or slaughter. Spores are transferred to vegetables by wind, rain, insects and handling during harvest. If viable spores remain after heat processing, favorable conditions will allow their germination and subsequent toxin production. The toxin is heat labile (destroyed by heat), so most cases of botulism are caused by foods that are not reheated after the initial processing, such as canned beans used in cold salads, or smoked fish.

There are about nine outbreaks and twenty-four cases of food-borne botulism annually in the U.S., fifty-four percent of which occur in five western states: California, Washington, Oregon, Colorado and Alaska. Foods subject to botulism include plant foods, fish, fish eggs, mushrooms, soups, and various sauces, but it is most frequently associated with improperly home-canned foods such as meat, fish and vegetables like beans or corn. These types of vegetables are low in acid content and therefore more accommodating to the growth of *C. botulinum* than are those higher in acid, such as tomatoes. However, tomatoes that have been bred to be sweeter are also subject to contamination with *C. botulinum*. Home-canned products that are not heated properly to destroy all spores are at risk, with seventy percent of botulism caused by such products. Home canning should done using a pressure cooker at 250°F for over fifteen minutes. Jars and cans must then be properly sealed.

The majority of the remaining thirty percent of botulism outbreaks are attributed to commercially canned foods. All processing parameters for canned goods in the United States are designed to destroy spores of this organism, so if botulism occurs as a result of a commercially-processed canned product, it is due to a breakdown in the heat treatment somewhere along the line. Rarely, botulism outbreaks can be traced to other circumstances. In one such outbreak, restaurant personnel wrapped unwashed potatoes in aluminum foil, baked them, and stored them at room temperature. The foil supplied anaerobic conditions, the baking process caused spore germination, and storage at room temperature allowed cell proliferation. The potatoes were then used to make salad, so were not subjected to further heat treatment, which would have destroyed the toxin.

Other scenarios of botulism involve products packed in oil, such as garlic. Oil excludes oxygen, and when left at room temperature, *C. botulinum* can grow. Manufacture of such types of products now includes the addition of acid to inhibit this growth, though a similar situation may be present in thick cuts of fatty meat or fish. The fat acts similarly to the oil, preventing oxygen penetration to the interior. Another outbreak of botulism was caused by mushrooms that were sautéed in butter, then left on a grill under a pan lid. The butter and pan lid kept out air, and the grill was not hot enough to inhibit *C. botulinum* growth.

Vacuum-packed cold cuts and bacon can support growth of *C. botulinum* as well, without causing noticeable spoilage. This organism does not compete well with other microbes, but non-spore-forming bacterial competitors are destroyed by the heat of processing. Vacuum packaging then excludes oxygen. With no oxygen and no competitors, *C. botulinum* is free to proliferate under the proper circumstances.

Some meats and vegetables high in protein may develop obnoxious odors as *C. botulinum* grows. However, lower protein/higher acid foods show minimal signs of spoilage. For instance, canned beans and corn contaminated with types A and B will look and smell normal. Spores of *C. botulinum* types A and E have been shown to germinate and produce toxin in smoked fish. Type A, which is able to break down protein, produces noticeable spoilage but type E does not. The toxin itself is odorless, tasteless and invisible.

New convenience foods of particular concern with regard to botulism are prepared by a French method called *sous vide*, which means "under vacuum." In this process, foods are wrapped in high barrier films to exclude air, then cooked under vacuum to preserve freshness. Before consumption, the foods are simply reheated. The initial cooking does kill vegetative bacteria, but not spores. Again, in low-acid foods with no air and no competitors, *C. botulinum* has free rein.

Even though botulinum toxin has a tragic legacy in foods, it has been successfully used to treat many afflictions due to its ability to prevent muscle contraction. Commercially called "BoTox" or "Oculinum," it was the first biological toxin licensed for treatment of human disease. The Food and Drug Administration approved it in 1989 for treatment of such conditions as crossed eyes and lazy eye, both of which are caused by uncontrolled muscle contractions. Minute amounts of refined toxin are injected into eye muscles to relax them, thus correcting the problem. The results are only temporary, however, so injections must be repeated.

Other therapeutic uses for botulinum toxin currently under investigation include treatment of painful neck and shoulder muscle contractions, clenched jaw, and writer's cramp. Even facial tics and spasms caused by cerebral palsy may be controllable via use of botulinum toxin. It is also used for more common ailments such as migraines, chronic back pain, stroke, and elimination of facial wrinkles, though it is not FDA approved for these applications.

Diagnosis and Treatment

C. perfringens

C. perfringens food poisoning has very characteristic clinical and epidemiological patterns that are almost conclusive in diagnosing the illness. Confirmation requires finding greater than one million organisms of the same serotype per gram of implicated food and stools of infected individuals. Symptoms last about twenty-four hours. Illness is self-limiting and patients usually recover completely without medical intervention, though it may be necessary to replace fluids and electrolytes if diarrhea is severe.

C. botulinum

Untreated botulism is often fatal and probably substantially under-diagnosed since single cases may go unreported. It is also easily misdiagnosed, often being confused with other neuromuscular conditions such as atypical *Guillain-Barré syndrome* or tick paralysis.

Serum of a patient suspected to be suffering from botulism will be toxic to mice that have not been protected with type-specific antitoxin. This mouse assay takes one to two days to complete, however, so the physician must pick up on the characteristic symptoms early and administer antitoxin in a timely fashion. This will minimize subsequent nerve damage but will not reverse existing paralysis.

Botulism antitoxin is maintained by the Centers for Disease Control and Prevention, and can reduce the severity of symptoms while nerve synapses recover over a time period of weeks to months. During this recovery period, supportive care, such as mechanical ventilation and nutritional support, may be necessary. Those who survive an episode of botulism may have fatigue and shortness of breath for years.

Antibiotics are not useful in treating botulism, since it is caused by pre-formed toxin in food and not the bacterium itself. Suspicious foods should be examined for toxin and/or *C. botulinum* bacteria.

United States law requires that botulism be reported to public health authorities, who in turn must contact the CDC. If the botulism was caused by commercially prepared foods, the USDA or FDA must also be notified. A suspected botulism outbreak requires a search for additional victims, confirmation of the diagnosis, and identification of the source of exposure.

Polymerase chain reaction (PCR) assays to detect *C. botulinum* DNA are under development, as are Enzyme Linked ImmunoSorbent Assays (ELISAs) to detect preformed toxin. Currently, laboratory assays for botulism are available only at the CDC and in twenty state and public health laboratories. Other diagnostic procedures are available to determine the cause of paralysis, including nerve stimulation studies and analysis of cerebrospinal fluid, but would neither be practical nor readily available in case of a large outbreak of botulism.

Clostridium Toxins as Biological Weapons

Clostridium perfringens

Researchers have discovered that toxins produced by certain foodborne pathogens such as *Staphylococcus aureus* and *Clostridium perfringens* are capable of triggering a massive immune response in their host, and have therefore been classified as "super-antigens." Apart from food poisoning, these toxins can cause inflammation of the brain and spinal cord, as well as a variety of autoimmune diseases.

C. perfringens produces a variety of toxins, the most potent of which is called "alpha toxin." It is able to break down lecithin, a crucial constituent of cell membranes, and would be lethal if delivered as an aerosol. In addition to typically acquired *C. perfringens* symptoms, those of inhaled alpha toxin may include an inability to breathe, anemia and elevated liver enzymes. Toxin absorbed by the skin would cause leakage and destruction of blood vessels and liver damage. Military experts say water-borne delivery is also possible, but less likely.

Direct contamination of foods could cause widespread toxicoinfection among the population. The resulting diarrhea would be more debilitating than lethal, except among immunocompromised populations such as the elderly or AIDS patients.

Clostridium botulinum

Botulinum toxin, along with anthrax, plague and smallpox, are among the most likely agents to be used for bioterrorism. Botulinum toxin is available to terrorists possessing the technical capacity to grow *C. botulinum* and harvest and purify the toxin, but toxin supplies for medical therapy are not potent enough for use by terrorists.

Development of botulinum toxin as a possible weapon began sixty or more years ago when prisoners of Japan's occupation of Manchuria were fed cultures of *C. botulinum*, with lethal effects. The U.S. first produced botulinum toxin during World War II as part of the biological weapons program. Per executive order of then President Nixon, 1969 brought an end to this program in the U.S.

In 1972, the United States, the Soviet Union, and more than one hundred other nations signed the Biological and Toxin Weapons Convention, which prohibited research and production of biological weapons. Never the less, Iraq and the Soviet Union later produced botulinum toxin and experimented with inserting the gene for its production into other bacteria. In 1988, a report by the Armed Forces Medical Intelligence Center stated that scientists in Baghdad had already produced weapons using *C. botulinum.* Russian scientists, formerly employed by the Soviet Union's defunct bioweapons program, have now been recruited by other countries believed to be developing botulinum toxin as a weapon.

In March of 1995, the Japanese group Aum Shinrikyo (Supreme Truth) released the nerve gas sarin into Tokyo subways. It was later discovered that this same group had experimented with botulinum toxin as a biological weapon. Luckily these attacks were not successful, most likely due to faulty microbiological techniques or equipment failures. The organisms were obtained directly from Japanese soil.

Botulinum toxins are the perfect weapons because they are extremely potent, lethal, easily produced and transported, and patients require prolonged care. An outbreak constitutes a medical and a public health emergency. Ordinarily, botulism is not contagious, but according to the Working Group on Civilian Biodefense, a toxin-producing microbe could be genetically modified to be contagious.

If toxins were delivered as aerosols, symptoms would most likely resemble those of ingested toxin, but without the initial abdominal cramps, nausea and vomiting, which are caused by bacterial metabolites other than the neurotoxin. According to The Working Group on Civilian Biodefense, one gram of this substance, evenly dispersed and inhaled, would kill over a million people. The Group also states that technical factors, such as the instability of the toxin in air, would make such dissemination difficult.

Botulinum toxin readily dissolves in water, however. But contamination of a municipal water supply would be difficult because the toxin is inactivated by standard water treatments such as chlorination, and an extremely large amount of toxin would be required to effectively contaminate large reservoirs. The toxin would remain stable for several days in untreated water, other beverages or foods. Water-borne botulism has never been reported.

Sabotage of food or water could produce an outbreak from a single meal or widely distributed outbreaks that would need to be distinguished from naturally occurring food-borne botulism. Features that would suggest terrorist use of botulinum toxin include a large number of cases, use of an unusual toxin type, such as C, D, F, or G, a common geographic area, or multiple simultaneous outbreaks lacking a common source.

In the case of a large outbreak, the need for ventilators and other necessary care might quickly exceed current capacities. The United States is now working to develop a reserve of mechanical ventilators and personnel who know how to use them.

Preventing Botulism

There is a preventative vaccine for botulism that is typically administered to laboratory personnel and other at-risk workers who are working with *C. botulinum*. In order to confer protection, doses must be administered at zero, two and twelve weeks, followed by a booster at one year. Pre-exposure immunization is neither advisable nor available for the general public.

Botulinum toxin is destroyed in foods by heating to an internal temperature of 185°F for at least five minutes. Food with a peculiar odor, appearance or taste should not be eaten. Aerosolized toxin will be inactivated in about two days, but intentional release would probably not be recognized in time to prevent additional exposures. Risk of inhaling the toxin can be diminished by covering the mouth and nose. Clothing and skin should be washed with soap and water, and surfaces decontaminated with chlorine bleach solution.

Research is underway to develop rapid diagnostic testing to recognize a terrorist attack with botulism. Enzymatic assays could replace the mouse assay as the standard diagnostic tool. The U.S. military's Biological Integrated Detection System currently uses ELISA assays to rapidly detect agents in the battlefield.

Distribution of antitoxin takes time, due to limited availability. Some standard detoxification methods, such as administration of charcoal by health care personnel in the case of ingested botulinum toxin, can be used in the interim, but additional studies are needed to verify their effectiveness.

The Working Group suggests development of a human antibody to neutralize botulinum toxin (the current source of antibody is horses) but this will require a dedication of sufficient resources. Immunity would be incurred for long periods, and substantial quantities could be produced in vitro (in a laboratory test tube). Such an antitoxin would deter terrorist use of botulinum toxin, but until it is available, the population will remain vulnerable.

Clostridium				
Disease	Symptoms	Transmission	Prevention / Treatment	Food / Water Sabotage
Botulism (C. botulinum)	Blurred vision, difficulty swallowing, nausea / vomiting progressive flaccid paralysis	Improperly-processed low- acid foods	Careful home-canning etc. / Antitoxin	Possible / Possible–need large amount (aerosol delivery of toxin possible)
Infant botulism (C. botulinum)	"Floppy baby syndrome" weakness, loss of head control	Syrup, honey	Do not use syrup or honey on pacifiers / Antitoxin	No / No
Tetanus (C. tetani)	Lockjaw, contracted muscle paralysis	Deep puncture wounds	Vaccination once every ten years / Prompt vaccination	No / No
Toxico-infection (C. perfringens)	Stomach ache, diarrhea, gas	Temperature-abused foods (esp. high protein)	Careful handling of foods / None	Possible / Possible (alpha toxin as aerosol possible)

CHAPTER FOUR

Brucellosis and Q Fever

Brucella species and *Coxiella burnetii*, the causative agent of Q fever, have a number of similarities, both as bacterial species and as agents of disease. The illnesses they cause can be transmitted from animals to man under natural conditions. Brucellosis occurs primarily among domestic animals.

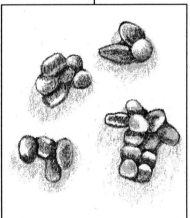

Brucella melitensis

Humans usually become infected by having close association with diseased animals, or by consuming contaminated raw milk or raw meat. Humans become infected with *C. burnetii* the same way, but while this organism is also a normal inhabitant of animals, it usually exhibits a parasitic rather than pathogenic existence, living within animal tissues without causing disease. In humans, both brucellosis and Q fever are systemic infections that have the potential to develop serious complications such as diseases of the bones, joints, heart, lungs, liver and nervous system. Another commonality is that both have undergone experimentation for use as biological weapons.

Historical Perspectives

Brucellosis

Bacteria of the genus *Brucella* have been known to cause disease in humans since 1887, when Sir David Bruce, a British bacteriologist, isolated the bacterium from the spleens of British soldiers who had died in the Crimean War of "Malta fever," so named because of its incidence on the island of Malta. The disease was also known as "Mediterranean gastric remittent fever." A Mediterranean Fever Commission was established to identify the causative agent and discovered that humans became infected with these bacteria by consuming unpasteurized goat's milk or cheese made from unpasteurized goat's milk. When the Commission restricted the intake of goat's milk, the incidence of this febrile (fever-associated) illness declined appreciably.

Later, it was discovered that other animals may also become infected with *Brucella*. *B. abortus*, which causes abortions in cattle, was discovered by a Danish veterinarian, B.L. Bang, in 1894. Brucellosis of cattle is thus also known as "Bang's disease." In humans, brucellosis is characterized by a waxing and waning fever, and so the disease has also been dubbed "undulant fever."

Q Fever

Q Fever first appeared in Queensland, Australia, in 1933, and in the United States in 1938. One might think the Q stands for the bacterium's point of origin, but it does not. The cause of this illness was initially unknown, so a question mark was often used on medical reports. This eventually led to the naming of the disease "Query fever" or "Q fever." *Coxiella burnetii* was identified as the agent in 1937, and named for H.R. Cox and F.M. Burnett, two of its early investigators.

There have been several significant outbreaks of Q fever among employees of meat-packing plants associated with the handling of raw meat. *C. burnetii* has the pathogenic potential to cause severe pneumonia.

Characteristics of the Organisms

Brucella species

These bacteria are *fastidious*, meaning that they require a rich environment for growth. These short bacilli find such an environment living as intracellular parasites within the mammary tissue of cows

(*B. abortus*), goats (*B. melitensis*), hogs (*B. suis*), and canines (*B. canis*). *B. abortus* and *B. suis* inhabit either cows or hogs, and are transferred between these animals if they share common feedlots or yards. *B. melitensis* is also routinely found in sheep and camels, and sometimes in bison, elk, caribou, and certain species of deer. Stray dogs and coyotes are more likely to carry *B. canis* than domestic dogs. While all four species are pathogenic to humans, *B. canis* infections are extremely rare.

Brucellae that are usually parasitic become pathogenic during animal pregnancy, when they invade the mother's uterus and cause abortion of the fetus. The bacteria are then shed in the infected animal's milk and urine. Other animals become infected when they ingest urine-contaminated feed. Humans contract undulant fever through direct contact with an infected animal, or by consuming contaminated raw milk or raw meat. Brucellae bacteria are quite resilient, and can persist in pond water and streams, grasses, dairy barns, and soil for prolonged periods of time—weeks or even months—thereby prolonging exposure time and increasing the number of ways the bacterium can be spread to humans and other animals.

Since its inception in 1956, efforts of the Brucellosis Eradication Program have been aimed at eliminating the disease among cattle in the U.S. The incidence of brucellosis among animals has been reduced by removing infected animals from herds and vaccinating young animals. By 1998, the known number of affected herds had decreased from 124,000 to just fifteen. Most remaining cases of animal brucellosis occur in California, Florida, Texas, and Virginia.

Coxiella burnetii

This bacterial species is similar to *Brucella* in its shape (short bacilli) and its lifestyle as an intracellular parasite. *C. burnetii* normally carries out a parasitic existence in sheep, goats and cattle, but in placental rather than mammary tissue. It ordinarily does not cause disease symptoms in these animal hosts, but is occasionally responsible for spontaneous abortions in goats and sheep. *C. burnetii* is shed in the milk, urine and feces of its carriers, and is present in particularly high numbers in the placenta and amniotic fluids, which readily contaminate the farm environment when an infected animal gives birth.

C. burnetii can also live in other vertebrates, particularly mammals. It does not form typical bacterial spores as do *Clostridum* or *Bacillus* species, but during replication within cells of its host, a large cell variant and a spore-like small cell variant are formed. It also exhibits two distinct antigenic states, Phase I and Phase II, that relate to its cell coating and pathogenic, antigenic, and immunogenic properties. The highly infectious Phase I is typically isolated from infected animals, while Phase II is non-infectious.

C. burnetii is the only food-borne member of the order *Rickettsiales*. The most famous member of this bacterial order causes Rocky Mountain spotted fever, which is spread by arthropod (fleas, lice, ticks) bites, as are other rickettsiae (*C. burnetii* can also be transmitted by arthropod bites). Also in contrast to other rickettsiae, which exist inside cells of a vertebrate host, *C. burnetii* can survive extracellularly (outside of a host cell), even in the air, with the potential of causing illness by inhalation. It is also the most heat-resistant, non-spore-forming bacterial species, so modern milk pasteurization parameters are designed to eliminate it.

The Diseases: Undulant and Q Fevers

Modes of Infection

Undulant fever (human brucellosis) is a worldwide disease, occurring more frequently in Europe, Africa, Mexico, India, central Asia, central and South America, and the Middle East, though its incidence has been greatly decreased by pasteurization. Q fever is also worldwide except in Scandinavia. Most U.S. cases of undulant fever, numbering only one hundred reported per year, are due to direct contact with infected animals, and occur primarily among veterinarians or those in animal husbandry. Q fever and undulant fever have both been linked to individuals working with raw meat, such as those employed in meat packing plants.

C. burnetii and the various species of *Brucella* are very similar in the means by which they create human infection. Both are considered to be occupational hazards among veterinarians and farmers. Organisms contaminate the environment via animal feces, urine, and in the case of *Coxiella*, when infected animals give birth. Animal waste materials become dry and bacteria-laden aerosols can be inhaled, thereby causing illness via the lungs. Undulant fever can be contracted through skin abrasions, such as small cuts on hands or arms, hence the association with the handling of raw meat. Either disease can be acquired through direct bacterial contamination of the eyes, and *Brucella* and *C. burnetii* can both infect humans via the gastrointestinal tract from ingestion of contaminated raw milk or raw meat, but the incidence of gastrointestinal infection has been practically eliminated through pasteurization. However, U.S. travelers returning from countries that do not pasteurize milk have been infected with undulant fever. Undulant fever is not transmitted person-to-person, and Q fever only rarely.

Coxiella does have one unique characteristic, however. Unlike undulant fever, Q fever can be transferred via an intermediate host. While it is rare for a human to be infected this way, bloodsucking arthropods such as ticks, fleas or lice can transfer Q fever from one animal host to another. *C. burnetii* bacteria present in the blood of an infected animal enter the arthropod as it takes a blood meal. The bacteria then multiply within the arthropod and are shed in its feces. When the next victim is bitten, the bacteria are deposited into the wound either by the contaminated mouthparts or fecal material of the arthropod.

In 1999, Q fever became a notifiable disease in the United States, which means that the doctor making the diagnosis is required to report it to public health officials. But most countries do not require that cases be reported so its true incidence is unknown. Undulant fever is also a notifiable disease in the U.S., but studies indicate that it is underreported: for every case that is reported, an estimated twenty-five to thirty cases are not. Over the last ten years, about one hundred cases of undulant fever have been reported annually in the U.S., so an estimated 2,500 to 3,000 people may actually have been infected.

Not only are undulant and Q fever similar in how they are contracted, but also in their intracellular lifestyle, which includes proliferation within human host cells, and in the symptoms they produce. When *Brucella* organisms infect a human host, the bacteria are phagocytized (engulfed) but are not destroyed. Rather, they proliferate within the phagocyte, where they are protected both from antibiotics and the host's immune system. When the phagocyte becomes full of new bacteria, it bursts, releasing progeny which travel through the bloodstream to create a systemic infection, targeting primarily the liver, spleen, bone marrow and sometimes the lungs, and causing abscesses in these areas as well as in lymph tissue. Literally the entire body may become infected, but symptoms may be confused with those of other infections or diseases. During acute infection, the bacteria may also multiply outside of phagocytes within extracellular body fluids, but if the condition becomes chronic, this multiplication becomes strictly intracellular.

The undulating fever from which the disease gets its name reflects the lifecycle of bacteria within the host. As bacteria are released into the blood from the bursting phagocytes, the host's immune system responds and body temperature goes up, resulting in fever. This typically occurs in the afternoon. Fever breaks in the evening, accompanied by profuse sweating, and stays lower until a new batch of bacteria is released the following afternoon. Throughout the duration of the illness, this cycle repeats.

All four species of *Brucella* can create this undulating fever, but *B. melitensis* and *B. suis,* which are better able to survive within the host, are the more virulent species. *B. abortis* and *B. canis* cause a milder disease with fewer complications.

C. burnetii also causes an acute systemic infection, Q fever, but it primarily infects the lungs, resulting in pneumonia. This organism is not a typical rickettsia. Besides being the only food-borne one, it causes the only rickettsial illness that does not begin with a rash, and the only one that can establish an infection via the respiratory or gastrointestinal tract, or through direct contact with the eyes.

Humans usually get Q fever by inhaling aerosols, i.e., bacteria-laden dust, containing *C. burnetti.* The bacteria multiply in the lungs and are engulfed by phagocytes. Similarly to *Brucella,* the bacteria multiply within the phagocytes, and are carried through-out the body via the bloodstream to create a systemic illness, though the lungs continue to be the primary target of infection.

Manifestations of Disease

The incubation time for undulant fever is one to two months, while that for Q fever is just two to fourteen days. Q fever lasts only two days to two weeks, but undulant fever may persist for weeks to months or even years, with interspersed periods of apparent health. Symptoms of both diseases are rather nondescript and flu-like, and include fever, headache, chills, and muscle pain. Fifty percent of those infected with Q fever will not exhibit any symptoms at all, while up to fifty percent will develop pneumonia with chest pain and coughing.

The onset of undulant fever may either be sudden or insidious (gradual). In addition to those symptoms listed above, undulant fever patients suffer malaise, anorexia and subsequent weight loss, fatigue, depression and joint pain. Twenty percent will experience a cough, and seventy percent of adults will have gastrointestinal symptoms. There may also be enlargement of lymph nodes, liver and spleen. Undulant fever may persist as a systemic illness, with or without symptoms. Q fever can also cause abdominal pain, nausea, vomiting and diarrhea, and thirty to fifty percent of patients with symptomatic Q fever will develop pneumonia.

The death rate for Q fever is about one percent, but if the condition becomes chronic, it can be as high as sixty-five percent. Untreated infections with *B. melitensis* yield a six percent death rate, whereas undulant fever caused by other *Brucella* species results in death less than one percent of the time.

Complications

Q fever is usually self-limiting, and most patients will recover completely within several months, even without treatment. Chronic Q fever, defined as that which lasts longer than six months, can develop any time between one and twenty years after an initial infection. This can be life-threatening and may lead to a myriad of debilitating symptoms such as severe headaches and chills, prolonged weight loss, congestive heart failure, muscle and facial pain, speech impairment and even visual and auditory hallucinations. Rare complications include hemolytic anemia and glomerulonephritis (inflammation of renal capillaries).

Pulmonary disease is a complication of less than fifteen percent of undulant fever cases. Also rarely, hepatitis, meningitis, and arthritis can develop following episodes of either Q fever or undulant fever, but the most serious complication of both diseases is endocarditis, an inflammation of the lining of the heart, which may evolve months or years following acute infection. Only one to two percent of infected individuals will suffer endocarditis, usually those having preexisting heart valve problems, but this condition is responsible for the majority of deaths from these diseases. Both illnesses can also lead to a chronic fatigue-like syndrome that can be very debilitating and difficult to treat.

Diagnosis and Treatment

Diagnosis of either disease based on symptoms alone is difficult. Those of undulant fever are indistinguishable from those of tularemia or typhoid fever. Doctors diagnosing Q fever must rule out tularemia, as well as plague and other causes of pneumonia such as *Legionella, Mycoplasma, Chlamydia* and viruses. Approximately half of the individuals infected with *C. burnetii* will have chest X-rays showing abnormalities similar to those caused by a viral infection since the lungs are the primary sites of infection. Chest X-rays of undulant fever patients, however, are normal. Q fever patients will also have elevated liver enzymes.

Samples for diagnosis of undulant fever include patient nasal swabs, sputum, respiratory secretions, blood or bone marrow. These can be tested for the presence of *Brucella* species using bacterial culturing techniques and PCR. Serology (ELISA) can be done to detect antibodies to brucellae in the patient's blood, but such antibodies take two weeks to develop, and can cross-react with those formed during a tularemia or cholera infection.

For diagnosis of Q fever, nasal swabs, sputum, respiratory secretions and blood samples are used for PCR, but culturing of *C. burnetii* is not done because this organism is extremely difficult to grow in the lab, and is very dangerous to lab workers since it is highly infectious—a single inhaled bacterium can start an infection. Serology using ELISA or indirect immunofluorescence (IFA) assays can effectively be used to diagnose Q fever.

In IFA, antibodies specific for the invading organism are labeled with a compound that causes them to glow bright green under ultraviolet light. *C. burnetii* bacteria can also be detected with immunohistochemical staining, wherein antibodies to the organism are labeled with a stain. Antibodies stick to the bacteria, which are then visible as colored deposits when viewed in a microscope.

Undulant fever can be treated with a combination of antibiotics for a period of six weeks. Some of these combinations are also useful as preventative therapy. Q fever can also be treated with various antibiotics, some of which can be used following exposure to prevent symptoms or delay their onset. If endocarditis develops as a complication of either illness, it is much more difficult to treat, requiring up to four years of therapy using a combination of antibiotics. Some individuals with severe endocarditis may require surgery to replace damaged heart valves.

Preventative Measures

Naturally occurring undulant fever and Q fever can be prevented by washing clothing and hands after coming in contact with infected animals, and by properly disposing of animal wastes and afterbirth. Avoid touching the eyes with contaminated hands, or consuming raw milk and raw meats. Germicides are also effective in killing *Brucella* species, though *C. burnetii* is resistant to many disinfectants.

Currently there are no effective human vaccines available for undulant fever, even though animal vaccination programs have been very successful. A human vaccine for *B. suis* is under development. A Q fever vaccine for humans is available and licensed in Australia, a single dose of which confers immunity for five years or more to naturally occurring Q fever, and is ninety-five percent effective against illness caused by inhalation of aerosols. People that have been previously exposed to Q fever may experience severe reactions to the current vaccine, however, so a new version is under development. Also available for investigational use on at-risk persons, but not the general public, is an inactivated whole-cell vaccine. An animal vaccine for *C. burnetii* has been developed as well, but is not yet available in the U.S.

Potential Use of *Coxiella* and *Brucella* as Biological Weapons

The United States has been involved in the testing of *Brucella* and *Coxiella* and their potential for use in biological warfare. Q fever was evaluated in 1955 using closely monitored human volunteers, none of whom died. It was during this study that the extreme pathogenicity of this organism was discovered; just one inhaled bacterium was enough to cause illness. The study also showed the effectiveness of *C. burnetii* organisms delivered as an aerosol. When jet fighters were used as a means of dissemination, the aerosols traveled fifty miles. Due to the successful nature of these experiments, 120 gallons of *C. burnetii*, enough to incapacitate millions of people, were produced in 1956.

In the 1980s, the Russians tested not only Q fever and brucellosis, but also anthrax, glanders, smallpox, plague and tularemia. *Brucella melitensis* and *B. suis* were among the CIA's stockpile of biological weapons that were retained after Nixon's 1969 ban on offensive biological weapons.

The American Type Culture Collection, (referred to in Chapter Two) in 1982, sold and delivered a variety of bacterial cultures to the University of Baghdad in Iraq. The assortment included three types of *B. anthracis*, five types of *C. botulinum*, and three types of *Brucella*.

In Japan, a Tokyo subway system was poisoned in 1995 with sarin nerve gas by the cult Aum Shinrkyo. It was later discovered that this same cult had attempted to make biological weapons using Q fever, botulism and anthrax.

There are mixed opinions regarding the usefulness of *Brucella* species as biological weapons. Undulant fever is readily curable with antibiotics and fatalities are low. These organisms are readily killed with germicidal agents and with the heat of pasteurization or cooking. Yet the CDC reports that the probability for use of *Brucella* species as biological weapons is high, and plans to improve epidemiological and diagnostic protocols, especially rapid testing techniques. The U.S. military concurs, stating that these organisms may be produced in a biological weapons laboratory, using a single cell or a small amount of the organism found in nature.

Perhaps a more likely candidate is *C. burnetii*. This is a resilient species, resistant to drying, sunlight, harsh environmental conditions, and many disinfectants. It is also extremely pathogenic—a single inhaled bacterium can establish infection—and very likely to cause illness if inhaled, though little information is available regarding long-term ramifications of an intentional release. Most likely, the disease would show similarities to that which occurs naturally. Q fever is rarely fatal, but it could potentially be used to incapacitate troops.

Intentional Contamination of Food or Water

If intentionally released, *Brucella* or *C. burnetii* aerosols would presumably be able to contaminate and persist in water, which would have to be treated with iodine or boiled. Aerosols would likely infect livestock as well, so care would need to be taken to avoid human contamination, including thorough cooking of meat and pasteurization of milk.

While it is possible for a terrorist to purposely add either of these organisms to food or water supplies, the nature of Q fever and undulant fever as diseases, as well as the history of experimentation with these organisms as biological weapons, suggests that for maximum effectiveness, either would most likely be delivered as an aerosol. However, sabotage of food or water should be considered. Regardless of the method of dissemination, antibiotic therapy would keep death rates low.

Brucella & Coxiella

Disease	Symptoms	Transmission	Prevention / Treatment	Food / Water Sabotage
Brucellosis (Undulant fever) (*Brucella species*)	Headache, chills, myalgias, weight loss, fatigue, diarrhea, vomiting, afternoon fevers that break with profuse sweating	Direct contact with infected animal; ingestion of raw milk or raw meat; inhalation of contaminated dust etc; bacteria in eyes	Avoid contact with infected animals; do not eat raw milk or raw meat / Antibiotics	Possible but not likely / Possible but not likely (Probably would be delivered in aerosol form)
Q fever (*Coxiella burnetii*)	Fever, headache, chills, myalgias abdominal pain, vomiting, diarrhea, possibly pneumonia	Bite of infected arthropod (rare); inhalation of contaminated dust etc; ingestion of raw milk or raw meat; bacteria in eyes	Same as above	Same as above

CHAPTER FIVE

Yersinia: Plague

Microbes that once devastated entire populations no longer enter our thoughts, having been practically eradicated by immunizations, antibiotics and improvements in sanitation. At least they have not crossed our minds until now. The subjects of this chapter are two members of the genus *Yersinia*, home

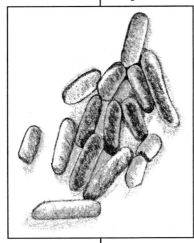

to a number of human pathogens, the most famous being that which causes bubonic (or pneumonic) plague, *Yersinia pestis*. No other pathogen in the world's history has had such an effect on human life. *Y. pestis* has been responsible for three pandemics (worldwide epidemics), the most famous of which, known as the Black Death, drastically reduced the European population in the Middle Ages, and had ramifications that extended far beyond loss of life, influencing the cultural, economic and political aspects of the time.

Yersinia pestis

Y. pestis tops the CDC's list of biological warfare agents to be feared: it is highly contagious, easily spread, and would likely kill one third of those it infected. Its close relative, *Y. enterocolitica*, is a worthy food-borne pathogen in its own right, possessing several nasty virulence traits, including the ability to invade mammalian cells.

The History of *Yersinia*

Yersinia pestis

The first record of plague tells of an epidemic in Athens in 430 B.C. The original pandemic raged from Egypt across Europe in 541 A.D., leaving fifty to sixty percent of the populations of North Africa, central and south Asia, and Europe dead in its wake. The bacterium responsible, *Yersinia pestis*, has taken its rightful place in history as the causative agent of this and two additional plague pandemics, and still causes epidemics in some developing countries today.

The second pandemic became known as the "Black Death" because the skin of victims took on a blackish cast as a result of decreased oxygen supplies to body extremities. It began in 1346 in Mongolia, when fleas infected with *Y. pestis* were carried on rats in search of food. These rats infested human dwellings, and the fleas they harbored transferred plague to human victims with their bites. Fleas also carried the disease across Asia on fur traders' pelts, travelers' blankets and clothing, and stow-away rodents.

Historical accounts reveal that the people of Europe were aware that the disease was approaching. Rumors stated that India, China and Asia Minor were literally covered with the bodies of its victims. In the year 1200, the population of China was 123 million. Following the aftermath of the plague, it had dropped to 65 million.

When plague did reach Europe, it wiped out a third of the population—twenty to thirty million people. Rioting, a drastic redistribution of wealth, unwarranted and misplaced persecution, as well as displacement of multitudes of people from their homeland affected art—paintings of the time often depicted death and the plague—politics, and religion. Christians believed that the plague was a call to repentance. As people struggled to understand the devastation they faced, blame was placed on Satan, evil spirits, and even the Jewish people, who were accused of poisoning water sources; they were the only population who took their drinking water from wells, which remained free of *Y. pestis*. This pandemic continued to spread from continent to continent, and over its 130-year duration, was responsible for the deaths of one hundred million people throughout the Middle East, Asia and Europe.

The third and final pandemic began in China in 1855 and spread across several continents. Ships visiting busy ports in China carried plague to the rest of the world, especially India. Over twelve million people died in India and China alone. During this outbreak in 1894, the causative agent was identified by A. Yersin, a Swiss-born French bacteriologist. He originally named the organism *Pasteurella pestis* to honor his teacher, Louis Pasteur, but the name that stuck, *Yersinia pestis*, was the one honoring its discoverer.

The most recent U.S. rat-borne epidemic was in the mid 1920s in Los Angeles. The last known case of plague to be transferred person-to-person in the U.S. occurred during that outbreak, though this type of transmission continues to be a problem in developing countries. Between 1947 and 1996, 390 cases of plague occurred in the U.S., eighty-four percent of which were bubonic plague. New Mexico, Arizona, California and Colorado accounted for the majority of these cases. In the 1980s, there were about eighteen cases of plague recorded in the U.S. annually. Most victims were younger than age twenty and one of every seven died. There were also epidemics each year throughout the 1980s in Africa, South America and Asia.

Elsewhere in the world, epidemics of a smaller scale have occurred as recently as 1997. In Madagascar, one patient effectively transmitted the disease to eighteen others, eight of whom died. Worldwide, there continues to be between 1,000 and 2,000 cases of plague each year, but future pandemics are unlikely thanks to better living conditions, public health initiatives, and the availability of antibiotics. Plague is still common among rodent populations on every populated continent except Australia.

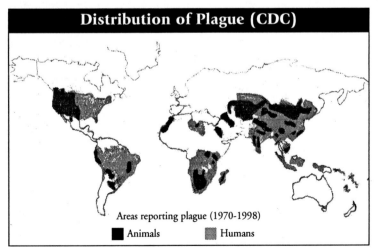

Distribution of Plague (CDC)

Areas reporting plague (1970-1998)

■ Animals   ▨ Humans

Yersinia enterocolitica

Yersinia enterocolitica was previously considered to be nonpathogenic but has now emerged throughout the world as a virulent pathogen in both animals and man. Indeed, most isolates are not pathogenic, but those strains found in pigs and pork products often are, and most outbreaks are due to contaminated pork.

As far as food-borne pathogens are concerned, *Y. enterocolitica* is relatively new on the scene, having only been recognized as a causative agent of gastroenteritis during the last forty years. This bacterium was first isolated in the state of New York in 1933 by M.B. Coleman, and recognized as an agent of gastroenteritis in 1939 by Coleman and his associate, Schleifstein. The illness, *yersiniosis*, is endemic to New York State, and is sometimes referred to as "Adirondack disease."

Another *Yersinia* species, *Y. pseudotuberculosis,* also causes gastroenteritis. This organism has not been reported as an agent in the U.S. Today it is found only Japan, where it has been transmitted in food and water.

Y. enterocolitica did not become widely recognized as causing illness until the 1960s, and by 1995, the World Health Organization had selected it as one of the most important emerging food-borne pathogens. The rise in incidence of yersiniosis follows the trend in developing countries to raise large quantities of swine in close quarters, where infection is easily spread between animals. In the Netherlands, Australia, Canada and Belgium, *Y. enterocolitica* has been isolated from victims of acute bacterial gastroenteritis more frequently over the past twenty years than have *Salmonella* (Ch. 2), *Shigella* (Ch. 8) and *Campylobacter* (discussed in Ch. 2). It has traditionally been more prevalent in Europe than in the U.S., particularly in Belgium, simply due to the fact that pork is eaten more frequently, sometimes raw. The organism is killed by heat, but the handling of raw pork has led to cross-contamination of other foods, such as fresh vegetables, that are eaten without being cooked.

Characteristics of the Genus *Yersinia*

Members of this genus are short, rod-shaped bacteria that are able to live with or without oxygen. *Y. pestis,* typically carried by rat fleas, has a variety of virulence factors that enable it to survive inside a human host, including the abilities to utilize nutrients, avoid phagocytosis and other mechanisms of defense, and damage host cells. *Y. enterocolitica* has some nasty virulence properties of its own, such as the production of a heat-stable enterotoxin, and the ability to bind and invade intestinal epithelial cells. *Y. pestis* is motile (capable of movement), while its food-borne counterpart, *Y. enterocolitica,* is motile only at temperatures at or below 86ºF.

Y. enterocolitica has an unusual propensity for growth at cold temperatures. While actually preferring to live at 82ºF, these organisms can grow at much colder temperatures, making them problematic as contaminants of refrigerated meats. It is even resistant to freezing. It is found in a wide variety of healthy animals in mesenteric (intestinal) lymph nodes and the intestinal tract. *Y. enterocolitica* has been isolated from the beginning of the large intestine of domestic animals, and fecal material of a wide variety of domestic and wild animals, including cats, dogs, birds, guinea pigs, horses, raccoons, chickens, chinchillas, beavers, rats, camels, cattle, deer, lambs, oysters, fish, and swine, primarily on the tonsils. It is not part of the normal flora of humans, but is sometimes isolated from clinical laboratory specimens from wounds, feces, sputum or mesenteric lymph nodes. Environmentally, it occurs in soil and surface waters, particularly in cool months, most likely due to waste contamination from wild animals. It has also been found in well water.

Y. enterocolitica is present in a wide variety of foods as well. This organism has been isolated from cakes, beef, lamb, mollusks, and ice cream, and is often found on fresh and vacuum-packaged meats of all kinds. Outbreaks have been associated with raw, pasteurized and powdered milk, raw vegetables, soy products, seafood, poultry, and tofu, but pork is the single most important source of human infection. The bacterium is transmitted primarily in raw or undercooked pork products, but its presence is usually due to a lapse in food processing since it is killed by heat. The fact that *Y. enterocolitica* grows in the refrigerator, along with an absence of competitors due to cold storage temperatures and lack of oxygen, makes it a frequent contaminant of vacuum-packaged meats. Outbreaks have also been associated with bean sprouts washed in contaminated well water, and chocolate milk that was apparently contaminated after the milk was pasteurized, during addition of chocolate syrup. (*Y. enterocolitica* is killed by pasteurization.) Rarely has it been transferred via contaminated blood transfusions or poor personal hygiene, wherein fecal material from an infected individual ends up on someone's hands, and eventually in their mouth. This is known as the "fecal-oral" route of infection.

Manifestations of Disease

Plague

The various names used for the different kinds of plague reflect disease manifestations. *Bubonic plague* is transmitted by the bite of an infected rat flea. Periodic outbreaks of plague in rodent populations sometimes kill a large number of the organism's normal flea hosts, such as rats, so fleas exit rodent corpses in search of blood and end up infecting humans or other animals. Human plague epidemics in developing countries are often due to fleas carried by house rats, and outbreaks in the U.S. are due to fleas of wild rodents. Bites of infected fleas deposit perhaps thousands of bacteria into the skin. In response, the host's lymph nodes capture the organisms but fail to destroy them. The bacteria then multiply, killing massive numbers of cells in lymph node tissue and creating "buboes" and pus-filled boils. The organism eventually gains access to the blood and migrates to the liver, spleen, and brain, causing hemorrhagic destruction of these organs.

Symptoms of bubonic plague—sudden onset of fever, chills, and weakness—begin two to eight days after receiving a bite from a plague-carrying flea, and are followed about a day later by the development of characteristic buboes. Bubonic plague gets its name from these inflammatory, painful swellings of the lymph glands that occur all over the body, but particularly in the groin. Buboes are so painful and tender that patients may not be able to move the affected area of the body. Disease progresses rapidly as bacteria enter the bloodstream, creating a condition called plague septicemia.

A few patients will get *primary septicemic plague* from a fleabite. In this type of plague, buboes are not formed, but as the name suggests, infection stills spreads throughout the body. Primary septicemic plague and bubonic plague can both lead to *secondary pneumonic plague*, also known as plague pneumonia, which results when bacteria migrate to the lungs. Symptoms of secondary pneumonic plague are chest pain, cough, shortness of breath and hemoptysis (the coughing up of blood). Plague pneumonia is fatal in fifty percent of cases.

Over the last fifty years in the U.S., about twelve percent of plague cases advanced to the latter stage. Neither bubonic nor primary septicemic plague are spread person-to-person, but secondary pneumonic plague can be contracted by inhaling respiratory droplets from an infected individual, and is spread via coughing. Plague acquired this way results in the fourth type, *primary pneumonic plague.*

Primary pneumonic plague occurs only rarely in the U.S., and represents only two percent of total cases. Pet cats and sometimes dogs may become infected with plague if bitten by infected fleas, or if they eat infected rodents. Patients of two recent U.S. cases contracted plague by inhaling respiratory droplets of infected cats. These patients exhibited not only characteristic lung symptoms, but also gastrointestinal symptoms—nausea, vomiting, abdominal pain and diarrhea. Treatment was delayed more than twenty-four hours and both patients died.

Most naturally occurring plague is due to flea bites, but plague bacteria will be present in the blood, tissue and body fluids of infected animals and can be contracted by humans through breaks in the skin. This method of infection occurs infrequently, usually as a result of skinning wild rabbits or other animals.

Occasionally, plague bacteria can set up housekeeping in the meninges (tissue covering the brain and spinal column) or cervical lymph nodes, leading to meningitis or plague pharyngitis (inflammation of the pharynx). Pharyngitis can also follow the inhalation or ingestion of plague bacilli. Toxins produced by *Y. pestis* act in combination with septicemia and lead to shock, coma and death. Gangrene of the extremities, such as toes, fingers, and nose, may occur in advanced stages of plague.

Yersiniosis

Yersiniosis is not a common illness. There is only about one confirmed infection per 100,000 persons annually in the U.S., or about 17,000 total each year. There are more in Europe, Scandanavia and Japan. Most yersiniosis outbreaks occur in October and November, while the fewest happen in the spring. The illness is generally relatively mild and occurs sporadically, making it difficult to establish its actual frequency. This organism is quite particular in its selection of victims, favoring the very young (less than one year) and the very old. Young children will experience fever, abdominal pain and diarrhea, which is often bloody due to the invasive nature of *Y. enterocolitica*. Older children and adults will have fever, and acute pain in the right lower abdomen that mimics appendicitis. On several occasions, yersiniosis victims have needlessly had appendectomies.

Other virulence factors include production of a heat-stable enterotoxin, and the presence of plasmids (small, circular bits of non-chromosomal DNA within the bacterial cells) that carry genetic information conferring the bacterium's ability to adhere to intestinal epithelial cells. Following adherence in the intestine, the bacteria enter underlying lymphatic tissue and establish an infection within monocytes (a type of white blood cell). If bacteria invade intestinal tissue, yersiniosis can progress to a highly fatal typhoid-like septicemia. *Y. enterocolitica* organisms have been recovered from the urine, eyes, cerebrospinal fluid, blood, and stools of infected individuals.

Yersiniosis is characterized by three distinct phases. Stage one, the acute phase, mimics appendicitis. Symptoms of stage one ensue two to three days after ingestion of contaminated food or water, and include abdominal pain, headache, diarrhea, and nausea. This self-limiting illness often progresses no further, and symptoms are gone in two or three days. Stage two may start two or three weeks later as the infection progresses to include inflammation of several areas of the body, including the heart muscle, skin, and gastrointestinal tract. Connective tissue or the nervous system may become involved as well, and arthritis may develop and affect the spine, causing a meningitis-like inflammation. This stage may require hospitalization of the patient.

Chronic conditions, including inflammation of the skin and muscles, and rheumatoid arthritis, comprise the third, *recurrent* stage of yersiniosis. Typically only those over the age of sixty will progress through all three stages. Stage one will affect infants and young children, and people twenty to sixty years of age will progress only to stage two.

A new mode of *Y. enterocolitica* infection was discovered in 1999. A CDC investigation of seven cases of *Y. enterocolitica* sepsis, spread randomly throughout the U.S., identified the source of the bacterium as blood from donors who had recently recovered from a gastrointestinal bout of yersiniosis, some as many as four weeks prior to giving blood. Because of its cold tolerance, the organism was able to grow very well in the nutrient-rich, refrigerated blood. Five of the seven transfusion recipients died.

Diagnosis and Treatment

Plague

Naturally occurring plague can be diagnosed by the incidence of characteristic buboes. Fever, exhaustion, headache, and the development of painful, swollen lymph nodes are also indicative.

Various rapid tests for plague do exist, including PCR, but these are only available at some state health departments, the CDC, and military laboratories. Tests for antibody formation in patient's blood are of little value, since antibodies are not formed for at least several days after onset of the disease.

Laboratory staining of sputum, lymph node aspirates or blood samples can be done to detect the unique shape and staining characteristics of *Y. pestis*, and direct fluorescent antibody testing may be positive. Culturing of the same types of samples can also be used to detect the presence of the organism. Incubation for twenty-four to forty-eight hours will allow the organism to grow, but unless a lab is equipped with automated or semi-automated identification systems, up to six days are required for definitive identification of isolates. The CDC and the Association of Public Health Laboratories are working together to establish criteria for laboratory diagnosis of plague, and measures for training lab personnel.

In the past, plague has been treated, per FDA approval, with streptomycin, which has reduced the death rate to between five and fourteen percent when administered early. FDA has also approved the use of tetracycline (though one study of *Y. pestis* strains from Madagascar indicated that thirteen percent of strains had some resistance to tetracycline), and doxycycline for prophylaxis, as well as treatment. The CDC recommends that individuals in contact with a diagnosed plague patient, especially of the pneumonic type, should begin preventative antibiotic treatment depending upon the degree and timing of the contact.

In addition to antibiotic therapy, pneumonic plague patients will require substantial supportive care since systemic infection will involve complications resulting from sepsis such as respiratory distress, disseminated intravascular (within veins and arteries) coagulation, multiorgan failure, and shock.

Yersiniosis

Y. enterocolitica can be isolated from stools or vomitus of infected individuals, or from other sites such as the throat, lymph nodes, joint fluid, urine, bile, and blood. It can be identified as the infectious agent by biochemical and serological tests of these samples, and by finding the organism in the suspected food. The organism can be presumptively identified in thirty-six to forty-eight hours using selective plating media and incubation at refrigeration temperatures (cold enrichment), but definitive diagnosis takes up to three weeks or more. Pathogenicity of isolates must be confirmed by the presence of virulence factors since many *Y. enterocolitica* strains are nonpathogenic.

Yersiniosis is usually self-limiting, and symptoms will be gone within two or three days, although they can last longer depending on the initial dose and the age and health of the victim. If the disease progresses beyond the first stage, antibiotics may be administered, though in 1988, it was discovered that some strains of *Y. enteroclitica* infecting humans via pork are resistant to chloramphenicol, most likely due to use of this antibiotic in pigs.

Preventing *Yersinia* Infections

Plague

Naturally-occurring plague is successfully controlled in developed countries by monitoring and controlling urban and rural rat populations. This involves keeping a close watch on human and rodent plague infections, and controlling rodent fleas with insecticides when rodent outbreaks do occur. This goal has not yet been achieved in other countries, so plague continues to be a threat.

In areas of the U.S. where plague among wild rodent populations is widespread, such as the Southwest where rock squirrels and their fleas are a frequent source, people should eliminate potential shelter for rodents in and around homes, work places and recreational areas. This includes brush, rock piles, and junkyards. Rodents should be kept out of buildings, and potential food sources, such as pet foods, should be inaccessible. Sick or dead rodents should be reported to health authorities, and pets should be kept free of fleas, especially during wild animal outbreaks. Preventative antibiotic therapy can be used if someone is exposed to ill rodents or is bitten by rodent fleas.

In the Pacific states, the primary carrier of plague is the California ground squirrel and its fleas. Other carriers include prairie dogs, chipmunks, other types of ground squirrels, wood rats, and less frequently, deer mice, voles, wild rabbits and wild carnivores. Domestic cats and dogs may either become infected with plague or bring infected fleas into the house.

A vaccine for plague has been used in the U.S. for those considered to be at risk, including laboratory personnel and those working in plague-affected areas. However, this vaccine does not provide protection against primary pneumonic plague. Research to develop such a vaccine is ongoing.

Yersiniosis

It has been suggested that *Y. enterocolitica* be added to the list of zero tolerance organisms for food, a list that also contains formidable pathogens like *C. botulinum* and *Salmonella*, but in the meantime, several steps can be taken to prevent yersiniosis. Avoid eating raw and undercooked pork, and wash hands thoroughly after handling it. Consume only pasteurized milk, cheese, etc. made from pasteurized milk. Wash with soap after contact with animals, and dispose of animal waste with care. To prevent the spread of food-poisoning organisms in the kitchen, be careful to clean cutting boards, utensils and countertops, and to avoid cross-contamination of raw and cooked foods. Raw seafood and unpurified water should not be consumed. Food processors must develop and enforce sanitation and processing parameters to ensure elimination of *Y. enterocolitica* in foods prior to refrigerated storage.

Y. pestis as a Biological Weapon

Several aspects of plague contribute to its high potential for use as a biological weapon: it is widely available, easily distributed as an aerosol, can be transmitted person-to-person so people not originally exposed to aerosol could also get sick, and potentially has a high fatality rate, with plague pneumonia usually killing at least fifty percent of its victims. It is highly infectious, particularly in its pneumonic form. *Y. pestis* also has a remarkable history as a biological weapon.

Apparently the first attempt at biological warfare was during the war between Christian Genoese sailors and Muslim Tartars (1346-1347 A.D.) in the port city of Caffa on the Black Sea. Corpses of members of the Tartar army who had succumbed to plague were catapulted over the city walls at the Genoese sailors, forcing the Genoans to flee back to Italy. It must be noted, however, that any disease that was successfully transmitted to the Genoans was probably due to rats rather than the flung corpses.

During World War II, the Japanese army established a secret biological warfare research unit in Manchuria. Based on their knowledge of the Manchurian plague epidemics of the early 1900s, the Japanese, particularly General Shiro Ishii, the physician leader of the research unit, became intrigued with the potential of *Y. pestis* as an agent of biological warfare for three reasons: disease outbreaks could be disguised as natural occurrences; the pneumonic form is contagious and could create death in numbers out of proportion to the amount of bacteria released; and the most virulent strains could be used to make a dangerous weapon. The Japanese experienced problems, however, when they tried to disseminate plague organisms using an exploding bomb. The heat and extreme air pressure generated by the explosion killed practically all the bacteria. This setback was overcome by using clay bombs to disperse human fleas that were infected with *Y. pestis*. Using this approach, bacterial survival reached eighty percent. On several occasions during the World War II, China purportedly was the target for the original flea bombs as well as mixtures of rice, wheat and fleas. These attacks succeeded in creating mini epidemics of plague.

The United States began working with *Y. pestis* in the 1950s and 1960s, followed by the Soviet Union in the 1970s and 1980s. Both nations improved upon the delivery system, eliminating the flea vectors and opting for aerosol dissemination. Also during the 1980s, U.S. scientists inserted the *Y. pseudotuberculosis* gene for human cell invasion into *Escherichia coli*, an otherwise harmless, normal inhabitant of the human gastrointestinal tract. Not to be outdone, in 1989, the Soviets improved upon *Y. pestis*, making it more resistant to heat, cold, vaccines and antibiotics, and added the gene for production of diphtheria toxin which causes fever, chills, headaches, nausea, and if not treated, heart failure. The U.S. was not able to produce quantities of *Y. pestis* prior to halting of the US offensive program, but the Soviet Union had already produced large quantities in the form of dry powders that were packed into warheads, artillery shells and bombs. As of 1998, dried *Y. pestis* was still being produced by the Soviet Union at the rate of 1,500 metric tons per year. While few U.S. scientists studied plague, thousands of scientists in the former Soviet Union reportedly worked with it.

In Ohio in 1995, a microbiologist was arrested after fraudulently obtaining a culture of *Y. pestis* in the mail, triggering new legislation on antiterrorism, but little information is available on groups or individuals trying to develop plague as a weapon.

According to experts, terrorists would most likely deliver plague as an aerosol. The size of the ensuing outbreak would depend upon the amount and virulence of the bacterial strain, environmental conditions at the time it was released, and the methods used for aerosolization. Symptoms would start one to six days following exposure, and would vary significantly from those of the naturally occurring illness. Inhalation of *Y. pestis* would give rise to primary pneumonic plague, initial symptoms resembling those of other severe respiratory illnesses. Fever, cough, difficult breathing, chest pain and other signs of severe pneumonia could be accompanied by bloody, watery, or pus-containing sputum, and possibly the gastrointestinal symptoms of nausea, vomiting, abdominal pain and diarrhea. Death would follow quickly, usually within two to four days.

According to the World Health Organization (WHO), about 110 pounds of such an aerosol containing *Y. pestis* released over a city of five million would have the potential to cause 150,000 cases of pneumonic plague and 36,000 deaths. The bacteria would remain air-borne for an hour and could travel as far as six miles. People fleeing the city would continue to spread disease beyond these limits. Currently there are no warning systems to detect such an aerosol, so the first indication would be ill persons presenting classic symptoms of pneumonia and septicemia (blood infection).

Potential for Food or Water Contamination

The most likely route of delivery for plague would be as an aerosol. Since *Y. pestis* does not form spores, it is sensitive to heat and sunlight and would not be able to persist in the environment for any length of time outside a host. According to the WHO, at most, an aerosol would be viable for one hour. *Y. pestis* may be able to survive for a time in the soil, but would not be a risk to humans.

There is a chance that it could be spread via municipal water supplies, though due to the effects of dilution, extremely large quantities would need to be used. Plague pharyngitis, associated with swelling of cervical lymph nodes, follows inhalation *or ingestion* of plague bacilli. This form of plague resembles tonsillitis, and occurs frequently among Southeast Asian women who kill fleas they have extracted from one another's hair by biting them.

Y. enterocolitica or *Y. pseudotuberculosis* have greater potential as food sabotage contaminants, but in their natural state, the diseases they cause are not normally fatal. More disconcerting is the use of their human cell-invasion gene, or those genes that enable these bacteria to persist in the soil and water, for genetic manipulation of other pathogens.

Yersinia

Disease	Symptoms	Transmission	Prevention / Treatment	Food / Water Sabotage
Plague (*Y. pestis*)	Fever, chills, weakness, formation of buboes	Bite of infected flea; inhaling droplets from infected person or animal; via skin abrasions when skinning infected animals (rare)	Control of rodent populations / Antibiotics & supportive care, if necessary	Possible but not likely / Not likely due to dilution effects (aerosol delivery more probable)
Yersiniosis (*Y. enterocolitica*)	Fever, nausea, diarrhea, appendicitis-like pain; bloody diarrhea in children	Variety of foods, especially pork, due to improper cooking or post-processing contamination	Avoid raw or undercooked pork / Replace fluids, antibiotics if necessary	Possible but not likely / Not likely due to dilution effects

CHAPTER SIX

Staphylococcus aureus: Toxins

Staphylococcus aureus is a common food-borne pathogen that has been responsible for countless outbreaks of food poisoning, particularly those associated with social functions like church picnics and family reunions. It produces several types of tox-

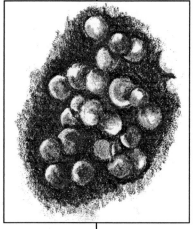

ins, but two in particular are of concern: Staphylococcus enterotoxin B (SEB) and toxic shock syndrome toxin (TSST). SEB makes us sick when we eat it, TSST when it gains entrance to the bloodstream through wounds or is absorbed through the skin, as in the case of the toxic shock syndrome (TSS) associated with tampon use. Both toxins can cause a very serious systemic poisoning and for that reason, have potential use as biological weapons.

Staphylococcus aureus

The History of *Staphylococcus aureus*

S. aureus first appears in history around 1870 as an agent of food poisoning. The illness it causes was originally referred to as "ptomaine poisoning," and sometimes still is, even though ptomaine can refer to any type of food poisoning.

In 1878, Robert Koch (see Chapter One, *Koch's Postulates*) and Louis Pasteur described the symptoms of food poisoning by *S. aureus*, and Koch associated the organism with the phenomenon of suppuration (pus formation). It is one of several bacteria that cause the human body to respond to infection by producing pus, which consists of lymph and white blood cells.

Medical and scientific literature dating back to 1908 refers to strains of *S. aureus* capable of causing an illness resembling scarlet fever, so named because of the sunburn-like rash characteristic of the illness. Historical accounts return once again to Queensland, Australia, (the originating point of Q fever), where in 1928, a pediatrician, while inoculating children against diphtheria, inadvertently poisoned them with *S. aureus,* which happened to be a contaminant in the diphtheria antitoxin he was using. Within twenty-four hours, eighteen of twenty-one children were ill, and twelve eventually died after suffering symptoms similar to those associated with scarlet fever. The children who survived developed skin abscesses at the inoculation site of the staph-tainted diphtheria antitoxin.

In 1914, on a farm in the Philippines, Dr. A.M. Barber became ill with nausea and diarrhea after drinking milk from one of the resident cows. He took some more milk back to his lab and drank it, but this time he did not get sick. He left the milk at room temperature for five hours then drank some more, and once again experienced symptoms of food poisoning. Further investigations by Dr. Barber revealed that the cow from which the milk came was suffering from mastitis, an infection of the mammary glands that is often caused by *Staphylococcus.* To see if these microbes were the culprits, Dr. Barber took sterile milk and inoculated it with staph, drank it, and quickly got sick. The rapid onset of illness led Dr. Barber to believe that a pre-formed toxin present in the milk was responsible, since not enough time had passed to allow the bacteria to become numerous.

This was proven to be the case in 1930 when researcher Gale Dack and associates at the University of Chicago had volunteers eat a cream-filled sponge cake that had been suspect in a food-poisoning incident. After volunteers became ill, *S. aureus* was isolated from the cake and grown in liquid media in the lab. The bacteria were then filtered out of the enrichment to yield a cell-free filtrate—one containing no bacteria, only the products of their metabolism. The plucky volunteers drank the filtrate only to get sick once more, proving that the responsible agent was a toxin produced by the bacteria and secreted into the environment—an exotoxin.

In 1980, toxic shock syndrome, a rapidly-progressing type of blood poisoning, made the news. Young, healthy women became devastatingly ill and sometimes died, apparently as a result of using super-absorbent, synthetic tampons. By 1983, the CDC had been notified of over 2,200 cases of TSS. Bad publicity resulted in the removal of certain brands of tampons from the market. Tampons that remain on the market today have information about TSS printed on the box.

Characteristics of Staphylococcus aureus

Staphylococcus aureus is an important human pathogen, albeit an opportunistic one. Man is the main reservoir of this organism, which is a normal inhabitant of fifty percent of the human population, residing on the skin and inside the nose, throat, eyes, intestinal tract and vagina of healthy individuals. When it has the chance, however, it can invade the body and is a common cause of infections in wounds, or more serious infections of the lungs, heart, blood and urinary tract.

Most domestic animals and non-human primates also carry *S. aureus*, and it is one agent of bovine mastitis. Milk cows with mastitis must be removed from production until the infection is eradicated, usually through treatment with antibiotics. Milk-receiving docks in milk bottling and cheese plants routinely test for the presence of antibiotics in the raw milk and if they are found, the entire tanker is rejected. Cheese cannot be produced with milk containing antibiotics because these drugs will kill starter bacteria. It is also important to reduce the incidence of antibiotic exposures in the human population, in attempts to control the problem of increasing antibiotic resistance seen among various bacteria, particularly staph.

S. aureus is a small, round (cocci) bacteria that under the microscope has the appearance of a cluster of grapes, hence its name: *staphyle* in Greek means "bunch of grapes." The species name

69

comes from the Latin word *aureus*, or golden, since *S. aureus* forms golden-colored colonies when grown on agar. This bacterium has the defining characteristic of being extremely salt tolerant, an important factor in determining which types of foods are subject to *S. aureus* contamination. Since it can grow in salt concentrations up to ten percent and can survive up to fifteen percent, it is often associated with cured hams and other salty meats. It is found everywhere in the environment—in air, sewage, water, milk, food, and even in dust, where it can survive for months at a time.

As an agent of disease, *S. aureus* is best known for its toxins, which are produced by thirty to fifty percent of strains. Some of these toxins, designated A thru E, cause gastroenteritis and thus are referred to as enterotoxins. Staph enterotoxins, especially SEB, also cause about fifty percent of the cases of toxic shock syndrome that are *not* associated with tampon use; the remaining fifty percent of toxic shock cases are caused by a particular type of toxin, TSST-1. Toxic shock syndrome is most commonly associated with tampon use, but can also result if *S. aureus* gains access to the body via surgical or other wounds, or as a complication of a severe skin abscess.

Staph food poisoning is a true intoxication. As was proven by Dack in the 1930's, staph food poisoning is not due to bacteria in the food, but rather caused by toxin that forms as the bacteria grow in the food. Staphylococci can grow at temperatures ranging from 50°F to 113°F, but toxin is only produced between 68°F and 113°F, and in the presence of oxygen. While the bacterium is killed by heat, the toxins are extremely heat stable, capable of withstanding thirty minutes of boiling. Staph toxins are also resistant to freezing, dehydration, irradiation, and stomach acid, and because they can trigger massive immune responses in humans, *S. aureus* is on the list of bacterial "superantigens."

Diseases Caused by *Staphylococcus aureus*

S. aureus is the causative agent of two distinct syndromes, food poisoning and toxic shock. The various toxins produced by this organism are responsible for both types of illnesses. Staphylococcal enterotoxins A through E produce a gastrointestinal illness and possibly toxic shock, whereas TSST causes only toxic shock.

Food Poisoning

S. aureus is responsible for about twenty-five percent of reported cases of food-borne illness in the U.S., although like so many other food-borne pathogens, its true incidence is not really known. Many people will not seek medical attention, and those who do may be misdiagnosed. The organism responsible for the illness may never be identified. Human carriers are usually the source of *S. aureus* contamination of foods. Handlers harbor the organism in their nose or in wounds on their hands or arms, and deposit bacteria into foods either directly from handling, or by coughing or sneezing. This emphasizes the need for food service personnel to use gloves, and not handle food at all if they are coughing or sneezing.

Most staphylococcal food poisoning is due to toxin type A, which is formed by *S. aureus* during the active phase of bacterial growth, when the bacterial population is doubling every few minutes. Toxins B and C are produced later on, after bacterial growth slows down. Toxins C and D are usually associated with milk and milk products. All staph enterotoxins cause irritations of the intestinal tract and symptoms of nausea, vomiting, cramping, and sometimes diarrhea. Severe intoxication also causes headaches, muscle cramps and changes in blood pressure accompanied with changes in pulse rate. Staph toxin is unusual in that it produces a signal that travels from the gastrointestinal tract through the vagus nerve to the brain, ultimately affecting the vomiting control center. Staph toxin is also very potent—less than one microgram of ingested enterotoxin is enough to cause illness.

Staph food poisoning is usually due to post-processing contamination, meaning the food has been cooked or baked and is contaminated after the heating process, usually by an infected food handler. Therefore, it occurs most commonly in food service and catering establishments where foods are handled a lot, and are subject to temperature abuse (not kept sufficiently hot or cold). *S. aureus* competes poorly with other bacteria, but these competitors are destroyed in the initial heating process.

Once *S. aureus* gains access to the food, it can grow if temperatures are appropriate, so hot foods should be held at 140ºF or higher, and foods to be refrigerated should be promptly cooled to 45ºF or less. Even if the food is reheated, the heat-stable toxin will not be destroyed. So the typical scenario for staph food poisoning is:

1) food is heated;
2) after heating, food becomes contaminated with
   S. aureus bacteria;
3) food is subsequently temperature abused, allowing
   the bacteria to grow and produce enterotoxins;
4) food is consumed.

The range of foods *S. aureus* can contaminate is quite broad: cooked meats, particularly ham and poultry, salads such as potato or macaroni salad, custards, cream fillings, milk, cheese, baked goods, eggs, fish, pasta, sandwich fillings like tuna, chicken or egg salad, creamed vegetables, and soups.

Staph food poisoning is characterized by a rapid onset of symptoms, usually one to four hours after eating food containing the preformed toxin. The onset time, as well as severity of illness, is dependent upon how much enterotoxin was present in the food, how much food was consumed, and the health and susceptibility of the victim to the toxin. These factors will also affect the duration of illness, but typically, it lasts no more than twenty-four to forty-eight hours, and is self-limiting. Patients usually recover to full health in two to four days, but if the illness is severe enough, may require fluid and electrolyte replacement. Patients are hospitalized in about ten percent of cases. Fatalities occur significantly less than one percent of the time, typically among the elderly, infants, or the severely debilitated.

Béarnaise sauce that had been held at room temperature was responsible for an outbreak at a restaurant. One of the twenty-six people that suffered staph intoxication became totally paralyzed. Recently it has also become evident that staph food poisoning victims can suffer from symptoms normally associated with toxic shock. If extremely high doses of toxin are ingested, it can cause blood pressure to drop to a dangerously low level, which in turn may lead to septic shock and death.

Toxic Shock Syndrome
From October 1979 to May 1980, fifty-five cases of toxic shock syndrome were reported to the CDC. Ninety-five percent of these were young women, thirteen percent of whom died. According to the last surveillance done in 1987, one or two women of every 100,000 in the U.S. between the ages of fifteen and forty-four will

get TSS each year, and five percent of them will die. TSS is a disease that affects many organ systems of the body at once:

- the gastrointestinal tract, producing vomiting or diarrhea;
- the musculature, causing severe muscle pain;
- mucous membranes of the vagina, the back of the throat, and the eyes, resulting in blood accumulation;
- the renal system, with elevated levels of blood urea nitrogen or creatine and urinary sediment;
- the hepatic system will evidence abnormal bilirubin or liver enzymes; and
- the central nervous system is affected with disorientation or unconsciousness.

Possible Sites of *S. aureus* infection

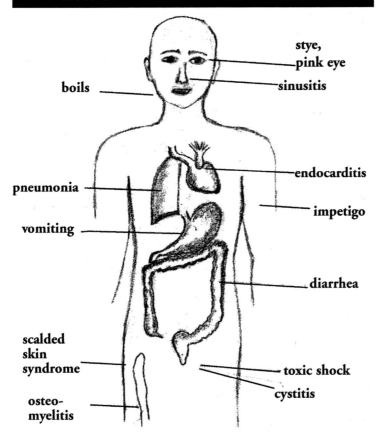

stye,
pink eye

boils

sinusitis

endocarditis

pneumonia

impetigo

vomiting

diarrhea

scalded
skin
syndrome

toxic shock

cystitis

osteo-
myelitis

Sometimes symptoms may be mild and flu-like, and at other times, can lead to death by heart, kidney, liver and respiratory failure.

TSS develops when *S. aureus* toxin reaches the bloodstream, either via a wound, complications of influenza, or absorption through vaginal walls as a result of using certain types of super-absorbent tampons or intra-vaginal contraceptive devices. Symptoms come on suddenly, and include a fever of at least 102°F, chills, muscle pain, headache, nausea, vomiting, diarrhea, and sometimes a sore throat. Within forty-eight hours, blood pressure may drop and the patient will become dizzy. A sunburn-like rash will develop, particularly on the palms of the hands and soles of the feet, and will peel in one to two weeks. Severe cases can be complicated by respiratory distress syndrome or cardiac dysfunction.

S. aureus seems to favor a viscous environment for growth, preferring foods such as thick creams and sauces, and protein-rich menstrual blood. Tampon use can lead to TSS by increasing surface area within the organism's normal habitat of the vagina, thereby giving the bacteria space and a good environment for growth and toxin production. Synthetic, super-absorbent tampons have more surface area than other types. Other factors contribute to make the environment ideal for staph, including hormonal changes that normally accompany menstruation, and an altering of the pH of the vagina, which is otherwise too low to allow proliferation of these organisms. The problem is compounded if the tampon is left in place too long, allowing yet more time for growth and toxin production, followed by absorption of the toxin through the vaginal wall and into the bloodstream where it proceeds to create systemic intoxication.

Other Illnesses Caused by *S. aureus*

S. aureus is a normal resident of human skin, carried by about fifty percent of individuals. It creates blemishes by entering skin surfaces at the base of hair follicles, where it establishes an infection. The body responds by sending an army of white blood cells to the infection site, resulting in the formation of a pus-filled pimple. If bacteria grow and spread from this point, abscesses or boils may form. If growth continues unabated, serious deep tissue infections can result, and may include abscesses in bone marrow, the spine or even the brain. *S. aureus* is also a causative agent of pink eye, an inflammation of the eye tissues, which is highly contagious, especially among children.

Respiratory flu, typically caused by a virus, can become more complicated if bacteria invade lung tissue or bronchial tubes that have been damaged by the initial viral infection. Potential bacterial invaders are *Haemophilus influenzae*, *Streptococcus pneumoniae*, or *S. aureus*. Bacterial lung infections can lead to fatal pneumonia.

Diagnosis and Treatment of Staphylococcal Illnesses

Food Poisoning

Clinical symptoms can be used initially to diagnose food poisoning due to staph, which is characterized by its short incubation period, sudden onset of symptoms and fairly brief duration. Epidemiological studies will usually reveal several victims, all having eaten the same foods at a function or restaurant. These victims should be interviewed to determine the food culprit, which then should be analyzed for the presence of staph in large numbers. Toxin levels necessary for illness are not obtained until the bacterial population exceeds 100,000 organisms per gram of food. The presence of the toxin itself can be ascertained using serological tests.

If viable bacteria are present in the food, or can be isolated from the stool of victims or from a suspected carrier, they can be identified using bacteriophage (or phage) typing. Bacteriophage are viruses that attack bacteria. The phage attach themselves to the bacterial cell wall and inject their genetic material into the bacterial cell. The viral DNA (or RNA) then "hijacks" the bacterial cell's replication machinery. Instead of more bacteria, new viruses are formed within the bacterial host cell, which eventually ruptures and releases the virus progeny. Phage typing can be used to identify bacteria because phage are very specific regarding the types of bacteria they will infect.

Toxic Shock Syndrome

TSS can also be diagnosed by its symptoms, especially if the victim has recently used super-absorbent, synthetic tampons. There is no antitoxin for TSST, so therapy is restricted to replacing fluids and other supportive care. Patients usually require intensive care. Antibiotic resistance among strains of *S. aureus* continues to be a problem in health care facilities.

Use of SEB or TSST as a Biological Weapon

Inhalational *S. aureus* Intoxication

As with many of the other potential biological weapons, *S. aureus* toxins could be released as aerosols and cause inhalational poisoning, which does not occur under ordinary circumstances. In most cases, this illness would only be debilitating, but could prove fatal if the dose were high enough. The manifestations would presumably mimic those of naturally occurring toxic shock syndrome, and either toxin, SEB or TSST-1, would result in similar scenarios. The incubation period would be as short as one hour or as long as twelve. Flu-like symptoms would come on suddenly and would include a fever of 102°F to 106°F lasting for several days, muscle pain, chills, headache, and a nonproductive cough (one that does not expel sputum) lasting up to four weeks, perhaps accompanied by a sunburn-like rash that would peel in one to two weeks, nausea, vomiting and diarrhea. Severity and duration of symptoms would most likely be dose-dependent. Severe cases may cause fluid to accumulate in the lungs, leading to respiratory distress syndrome and potentially respiratory failure.

Laboratory investigative studies suggest that the flu-like nature of this illness would render it easily confused with tularemia, plague, Q fever or inhaled botulism, so clinical symptoms would be of little value. However, anthrax, tularemia or pneumonic plague would continue to progress if untreated, while SEB intoxication would progress rapidly but eventually reach a state of stability. Studies also show that inhaled staph toxins have unique disease mechanisms and do not produce low blood pressure, as is usually seen in TSS. These inhaled toxins affect the respiratory system very quickly. Inhaled SEB also produces a high fever, which is not seen with the gastrointestinal illness. Serology could be used to rule out Rocky Mountain spotted fever, leptosporiosis (*Leptospira* infection) and measles.

Antibody levels in the patient's serum are not useful in diagnosis since a percentage of the population will have antibodies to staph toxins simply due to natural exposure. (These antibodies may actually be protective if the toxin dose is small enough.) Also, if a patient does have antibodies in their serum when they become exposed to staph toxin, the antibody levels will either be decreased or depleted by the presence of the toxin, and will be replenished as the patient convalesces.

SEB will not remain in a patient's blood serum for very long, but will accumulate in the victim's urine—high levels of SEB can actually inhibit kidney function—or may be present in respiratory secretions, and can be detected in either using PCR or other assays. PCR also provides a rapid method to detect the actual bacteria in blood samples, throat swabs, or cerebral spinal fluid aspirates. Blood samples will also show an elevated white blood cell count. White blood cells typically increase in number as the body tries to fight infection.

Treatment for inhaled SEB or TSST-1 would be limited to supportive care such as maintenance of fluid and electrolyte levels, rest, administration of pair relievers and cough suppressants, nausea control and, in severe cases, artificial ventilation. Since it is a toxin that creates illness, antibiotics are of no use, and an antitoxin is not yet available, though studies show that antitoxin would be beneficial for therapeutic as well as prophylactic use.

Historical Uses of S. *aureus* as a Weapon

In the late 1950s, Fidel Castro seized power in Cuba. In response, the U.S. made elaborate plans to oust him. These plans included the use of biological weapons, not for the purpose of killing people but rather to debilitate them, after which American troops would release a cocktail of microbes that would affect Cuban soldiers and civilians, rendering them incapacitated for an extended length of time and allow the Americans to overturn Castro.

Thousands of gallons of this cocktail were produced and tested but never used. It contained the virus for Venezuelan equine encephalitis, the bacterium for Q fever, and concentrated SEB. The latter was chosen because it had proven to be more effective than synthetic chemicals, requiring much lower quantities to elicit similar affects. The mixture was devised so that the various incubation times of these diseases would render victims helpless for up to two weeks. Within hours from the time of exposure, they would suffer symptoms of SEB intoxication—high fevers, chills, muscle pain, headache, nausea, vomiting, and diarrhea. In one to five days, victims would suffer the affects of the viral component—more nausea, diarrhea and a spiking fever—which would last one to three days and be followed by weeks of weakness and fatigue. Finally, after a long incubation period of ten to twenty days, Q fever would strike, causing two weeks worth of headaches, chills, fever, and possibly hallucinations.

Staph toxin was not originally among the biological weapons banned by the Nixon administration in 1969, but rather was reclassified as a chemical weapon. Eventually the renunciation was extended to include biological toxins. Ten grams of SEB were, however, contained in the lot of agents retained by the CIA. As recently as 1999, the U.S. was producing 1.9 metric tons of SEB each year.

In 1994 and 1995, at the U.S. Army Medical Research Institute of Infectious Diseases in Fort Detrick, Maryland, Rhesus monkeys were used in studies to determine the effects of inhaled SEB. Monkeys exposed to a lethal dose of the toxin for ten minutes exhibited gastrointestinal symptoms within twenty-four hours, but these were mild and ceased within forty hours. Lethal affects of the toxin came forty-eight hours after exposure with an abrupt onset of a rapidly progressing illness consisting of lethargy, difficult breathing, and facial pallor (paleness), which ended either in death or euthanasia of the monkeys within four hours. Lung and intestinal damage, hepatic lesions, swollen lymph nodes, and extreme immune system stimulation were discovered in autopsies. Staph toxins are now known to be extremely potent activators of T-cells, which contributes to *S. aureus'* status of "superantigen," and many of their effects are due to interactions with the host's own immune system.

In another enlightening but unfortunate incident, nine laboratory workers accidentally inhaled SEB and suffered varying degrees of illness having a three-to-four hour onset and three-to-four day duration. Three of the workers were classified as being seriously ill, two moderately ill, and four mildly ill. All nine victims experienced shaking chills as symptoms ensued, and all had fevers, some as dangerously high as 106°F, and all were hospitalized with a nonproductive cough. Eight of the nine had headaches ranging from mild to severe; seven had moderate pleurisy-like chest pains; six experienced nausea and anorexia and four of these also vomited for a short time. Five of the patients had difficulty breathing and exhibited abnormal breathing sounds. Only the three patients classified as having mild cases had normal chest X-rays, while the other six showed moderate compromise of the respiratory system. None of the victims had diarrhea and all nine recovered without the need for fluid replacement.

Aerosol Release of SEB

The consensus of experts seems to be that staph toxins delivered as aerosols would generally be incapacitating but not lethal, even though high doses could potentially cause death by pulmonary edema (accumulation of fluid in the lungs), though in less than one percent of cases. Most patients would be expected to recover quickly from the acute phase of illness, but if theses toxins were used on troops, personnel would probably not be able to resume duty for one to two weeks. The symptoms of inhalational staph poisoning would be similar to, though more severe, than naturally occurring food poisoning or TSS, and would exhibit higher rates of illness and death. Frighteningly, studies with mice have shown that the potency of staph toxin can be amplified by administering it along with endotoxin from other types of bacteria. Mice are not ordinarily affected by SEB; only humans, primates and kittens are susceptible to it. But when minute amounts of SEB were given to mice in combination with non-lethal amounts of endotoxin, the mice became ill.

SEB or TSST used as a weapon would probably not cause a lot of deaths, but, as in the case of Q fever, could be used to temporarily incapacitate troops, rendering as many as eighty percent of exposed individuals ill for a fairly long time, to the point that they would not be able to carry out their mission.

Protective Measures

Currently, antitoxin to SEB is not commercially available, but there is an experimental antitoxin that is capable of reducing the death rate if it is given early. Other preventative measures include thorough washing of hands and clothing, and surface decontamination with antimicrobials such as ten percent hypochlorite solution, which will destroy toxins in ten minutes.

Studies are ongoing to develop methods of immunity, both active (generated by the host's own immune system) and passive, wherein preformed antibodies are supplied. Experimental models show that passive immunotherapy must be administered within four to eight hours after inhalation to be of any use. An SEB vaccine has shown promise in animal trials, but has not been approved for human use. However, there are several human vaccines in development, which have been produced by genetic inactivation of the toxins.

Potential Contamination of Food or Water

Delivery of staph toxins as aerosols may not be practical due to the diluting affects of releasing the agent into the open air. Therefore, it may be more logical to deliver these toxins as food contaminants, or in small quantities of drinking water. In fact, experts have stated that staph toxin could be used this way in a special forces or terrorist mode.

Excessive amounts of toxin would have to be ingested to cause potentially fatal toxic shock syndrome. Most cases would be manifested as a self-limiting gastrointestinal illness, and ingested toxins can be neutralized with activated charcoal, administered by a health-care professional, if the patient is conscious and alert.

Staphylococcus aureus				
Disease	Symptoms	Transmission	Prevention / Treatment	Food / Water Sabotage
Intoxication (food poisoning) Toxins A - E	Vomiting, cramps, nausea, headache	Post-processing contamination of foods, usually by an infected handler	Do not handle foods if you are recovering from staph illness / Replace fluids & electrolytes	Possible / Possible (aerosol delivery of SEB or TSST also a possibility)
Toxic shock syndrome TSST–1 and SEB	Vomiting, diarrhea, myalgias, disorientation, fever, chills, headache, sunburn-like rash on palms & soles	Use of super-absorbent tampons, wounds contaminated with *S. aureus*	Change tampons frequently / Replace fluids & electrolytes, may require intensive care	Possible / Possible in small quantities of water only

CHAPTER SEVEN

Cryptosporidiosis, Cholera & Tularemia

The names of the three diseases discussed in this chapter illustrate the microbiologist's penchant for cumbersome nomenclature. Cholera may be the only one that sounds familiar, but all three have been around for quite some time, and all are transmitted by contaminated water and, less commonly, by contaminated foods. Cholera and tularemia are bacterial illnesses, but cryptosporidiosis, or "crypto" for short, is caused by a parasite, *Cryptosporidium parvum*.

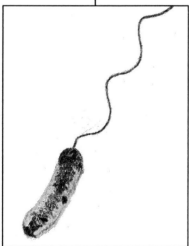

Vibrio cholerae

C. *parvum* was not very well known until 1993. That year, 403,000 people in the Milwaukee, Wisconsin area contracted crypto when this parasite inadvertently made its way into the municipal water supply, apparently due to heavy rains that created excessive runoff from farm fields laden with animal feces. During this infamous outbreak, one hundred people succumbed to this disease, which is particularly dangerous for AIDS and other patients with compromised immune systems, and is a significant cause of water-borne illness all over the world.

Cholera has always been well known—and feared. It is still common even today in certain parts of the world where it is perpetuated through poor sanitation and lack of proper water treatment. There have been seven cholera pandemics recorded, the most recent of which is still ongoing. Cholera is also on the list of emerging, or in this case *re*-emerging, infectious diseases—ones that have greatly increased in frequency over the last twenty years, and continue to increase in severity as we move into the future.

Although tularemia was once prominent, it is no longer a problem in the United States. Tularemia was removed from the list of notifiable diseases in 1994, most likely because the number of annually reported cases decreased substantially in the second half of the 1900s, along with a decrease in rabbit hunting, rabbits being a primary carrier of the disease. Less than two hundred cases have been reported each year since 1967. However, because of its high potential for use as a biological weapon, it was reinstated as a notifiable disease in 2000. It is extremely infectious and possesses the potential to cause illness and death.

Cryptosporidiosis
Historical Perspectives

Cryptosporidium parvum has, in all probability, existed for thousands of years in waters all over the world, but was not identified as an agent of human disease until 1976. Since that year, *C. parvum* has been linked to outbreaks of diarrheal illness in several countries, and is one of the most prevalent of pathogenic parasites—serological studies have shown that eighty percent of people in North America have been infected with it at one time or another, making it one of the most common causes of water-borne disease in humans in the U.S. Many outbreaks are due to contaminated drinking water, or accidental ingestion of contaminated water in water parks or community swimming pools. Outbreaks are also common in daycare centers.

The Organism

Members of the genus *Cryptosporidium* are intracellular, single-celled parasites. They inhabit the intestinal tract of humans and herd animals such as sheep, goats, cattle, deer and elk. There are several species, some of which are host-specific, but *C. parvum* is the only one known to cause illness in humans. It also infects young calves.

This organism has a rather complex life cycle that begins with the infective stage called an *oocyst*, which is about half the size of a human red blood cell. Oocysts are shed in the feces of an infected person or animal, and can contaminate water in lakes, streams or swimming pools. They can also be exhaled in respiratory secretions to contaminate the air. When ingested or inhaled by a new host, the oocysts rupture and release *sporozoites* that are able to penetrate the host's epithelial (skin) cells in the respiratory or gastrointestinal tract. Sexual reproduction takes place within these host cells to yield two types of oocysts: thick-walled, which are shed by the host (eighty percent of oocysts are of this type), and thin-walled, which continue to re-infect the current host. The exact disease mechanism is not known, but intracellular multiplication can result in severe tissue damage in the host.

The vegetative form of *C. parvum* is susceptible to chemicals, but the thick-walled oocysts are very resistant in the natural environment where they remain viable for several months under cool, moist conditions, especially in lakes, streams, ponds etc. They are also resistant to chemical disinfectants, and several outbreaks have indeed been caused via chlorinated swimming pools and water supplies. The oocysts can be destroyed by exposure to sunlight, drying, and boiling.

The Disease: Cryptosporidiosis
Similar to many of the pathogens already discussed, *C. parvum* is capable of producing illness in two ways: when the infective oocyst is ingested, an intestinal infection ensues; when oocyts are inhaled, this can lead to tracheal or pulmonary crypto. Naturally occurring crypto is almost exclusively of the ingested form. Large outbreaks, such as the one in Milwaukee, have been due to people swallowing contaminated water.

Crypto can also be contracted by ingesting contaminated food. Foods that have been implicated in reported outbreaks are usually raw fruits and vegetables that have been exposed to contaminated water, either through irrigation or washing, or that have been fertilized with manure containing oocysts from infected animals. Apple cider was the vehicle in one instance. Apparently the apples used to make the cider were rinsed with contaminated water. Even chicken salad has been a culprit. Pasteurized, canned or frozen foods do not typically serve as vehicles, but theoretically, as with staphylococcal enterotoxin B (SEB), any food has the potential to carry *C. parvum* if contaminated by an infected food handler. There are no current regulations to prevent *C. parvum* contamination of fresh foods.

Humans can also get cryptosporidiosis directly from animals, especially if they do not wash their hands after handling animals or animal fecal material. Person-to-person spread is possible via dirty hands or contaminated bathroom surfaces. Crypto may even be acquired in a hospital environment.

Crypto is also a common problem in daycare centers, where incidence of the disease can be high with outbreaks lasting up to four weeks. The illness is perpetuated by fecally-contaminated hands of workers, particularly those who change diapers, and children who fail to wash their hands after using the bathroom.

Most frequently, however, crypto outbreaks have been attributed to contaminated water supplies—swimming pools, hot tubs, fountains, lakes, rivers, springs, ponds and streams. Infected animals or humans contaminate these water sources with fecal material, depositing millions of oocysts at a time. Luckily the incidence in municipal water supplies is very low.

Symptoms of intestinal crypto start two to ten or more days after swallowing oocysts, and range from mild to severe, with watery diarrhea, stomach cramps and a low fever. Occasionally, victims may also have nausea, vomiting and weight loss. Symptoms of a crypto infection can become cyclic, and patients will alternate periods of apparent health with bouts of watery diarrhea. At the other end of the spectrum, crypto can also be asymptomatic (causing no symptoms in the infected individual). The severity of the illness is generally dependent upon the health of the victim.

Diarrheal episodes are usually self-limiting and will resolve in two weeks, although they can go on for several weeks and be quite serious, with as many as twenty-five bowel movements per day. A crypto infection can be life threatening to the immunocompromised, who may suffer as many as seventy stools per day with a loss of almost twenty liters of body fluids. AIDS patients may have this diarrheal illness for months or even years. Children, pregnant women, cancer patients, individuals with a deficiency in immunoglobulin A (IgA), or those taking corticosteroids are considered to be at risk and are more susceptible to dehydration.

If oocysts are inhaled, either from respiratory secretions or contaminated dust, pulmonary or tracheal crypto may result. This infection produces a cough and low-grade fever, and possibly intestinal distress. *C. parvum* is particularly virulent if inhaled, with less than ten, and possibly just one oocyst necessary to establish infection. The intracellular invasion and multiplication within cells of the host's pulmonary system can prove to be fatal due to excessive tissue damage caused by the organism. For the most part, however, pulmonary infections are limited to those with immunodeficiencies.

Diagnosis and Treatment

To diagnose crypto, oocysts can be identified in fecal material using a technique called acid-fast staining. Fecal samples are stained with a colored dye, then rinsed with acid alcohol. The dye is retained by the oocysts but is washed out of surrounding material by the acid alcohol. Biopsied intestinal, tracheal or pulmonary tissue can be examined using the same method. Fluorescent antibody detection can also be used. None of these assays are done on a routine basis at most laboratories, however, and because oocysts are difficult to identify using existing methods, new tests are being developed.

Currently there is no effective drug for the treatment of crypto, but research is also ongoing to find effective treatment therapies. Otherwise healthy individuals will recover on their own, and immunocompromised patients, per the advice of health care professionals, may use antidiarrheal or antiretroviral drugs, which will improve functioning of the immune system and either decrease or eliminate symptoms. These are not cures, however, and diarrhea may return when treatment is stopped. Patients taking immunosuppressive drugs should see their doctor if they believe they may have crypto. All victims of crypto should replace lost fluids and electrolytes.

Preventing Cryptosporidiosis

Since this illness is of particular concern to those with compromised immune function, the Centers for Disease Control and Prevention and the Environmental Protection Agency recommend that these individuals use bottled or purified drinking water. Boiling water for at least one minute will destroy oocysts. Faucet filters are also available that will remove the cysts, but they must be specially designed for this function. The absolute pore size must be less than one micron, or filters must be designated for cyst removal, effective for removing 99.95% of all particles in the size range of three to four microns, as well as oocysts.

Daycare workers should be careful to wash hands with soap after changing diapers or cleaning changing tables. Children should be reminded to wash their hands after using the bathroom. Anyone coming in contact with human or animal feces should wash their hands thoroughly.

Campers and other recreationalists should boil or filter drinking water taken from streams, lakes etc., but should not rely on chemical purifiers to do the job since *C. parvum* oocysts are resistant to disinfectants. If you are infected with crypto, do not go swimming until diarrhea has been gone for at least two weeks. *C. parvum* bacteria can be present in stool for several weeks after symptoms have ceased.

If a known municipal water outbreak is occurring, the best precaution is to use bottled drinking water. The same recommendation applies to those traveling to countries where the safety of drinking water is questionable. Remember that fresh fruits and vegetables and ice also are potential hazards. Rinse all fresh produce in clean water and peel it before eating it.

Intentional Contamination of Food or Water

Crypto is one potential biological weapon that is *more likely* to be used to sabotage food or water than it is to be dispersed as an aerosol. *C. parvum* oocysts are susceptible to environmental degradation by drying and UV sunlight, and would break down quickly if sprayed into the air, but are quite stable in water and certain foods.

Foods that would not be heated before being served could act as vehicles for *C. parvum.* Salad bars and other sources of fresh fruits and vegetables may be good candidates, but any food would suffice as long as it would not be heated or reheated prior to being consumed. A greater number of people would fall prey to an intentional contamination of a municipal water supply, which has proven to be an effective vehicle in the past due to accidental contaminations, as occurred in Milwaukee.

Crypto is generally not fatal, but anyone with a compromised immune system should take precautions to avoid this infection.

Honorable Mention: *Giardia lamblia*

Giarida lamblia is a flagellated (mobile) protozoan, considered to be the most common intestinal parasite in the world. It is a very common cause of water-borne diarrhea, especially in developing countries, but even in North America, it is the most frequently reported cause of nonbacterial diarrhea. Over 400 million people worldwide have been infected, as much as fifteen percent of the U.S. population, some without their knowledge since asymptomatic carriers are very common. Between 1971 and 1985, there were ninety-two reported outbreaks in the U.S. alone. The similarities between *G. lamblia* and *C. parvum* are quite striking.

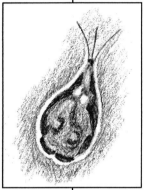

Giarida lamblia

The symptoms of giardiasis are identical to those produced by *C. parvum*, but the incubation period is longer—one to three weeks versus two to ten days—and may last for a year or more. The protozoan is transmitted to water and sometimes food by contamination with fecal material. Similar to crypto, the infective stage is a pear-shaped cyst that is resistant to chlorine. Beavers and muskrats commonly transmit this organism to lakes and rivers. Giradiasis is also highly contagious, with the infective dose somewhere around ten ingested cysts. It is also prevalent in daycare centers. Unlike crypto, giradiasis can be treated with drugs, but it can be very difficult to get rid of.

Giardia occurs in foods that have been washed with contaminated water, or touched by an infected handler. Lettuce, noodle salad, and raw fruits and vegetables are common vehicles. Ice cream, pudding, salmon, and tacos have also been implicated.

G. lamblia deserves honorable mention because it is a major problem for travelers, and its prevalence worldwide is cause for concern as an emerging pathogen. Given its availability and resistance to chlorine, it may very well be a prime candidate for sabotage of municipal water supplies.

Cholera

Historical Perspectives

This bacterial illness has a long history dating back to the 1800s, when it was prevalent even in the United States. The first of seven pandemics began in 1816. Cholera originally entered the U.S. by way of New York in 1832. It spread westward, reaching the upper

Mississippi River. It was introduced again in 1848, this time through New Orleans, and traveled north along the Mississippi River and as far west as California, carried by fortune seekers during the gold rush.

A cholera outbreak in London in the mid-1880s caused the deaths of five hundred people in a single week. A London physician, John Snow, traced the illnesses back to water from a single pump in downtown London. Dr. Snow's research, interviewing processes, and deductive reasoning, which he used in his attempt to identify the source of the illness, marked the beginning of the science of epidemiology, the study of disease incidence, control and distribution.

Cholera has practically been eradicated in the United States by modern sanitation and water purification practices. The last major U.S. outbreak was in 1911, and it was not seen again, except in travelers to foreign countries, until 1973. From 1973 to 1991, sporadic outbreaks were associated with consumption of raw, improperly cooked or cooked and recontaminated seafood, particularly shellfish, taken from the Gulf of Mexico. Apparently *V. cholera* had been reintroduced into these coastal waters, presumably by ships off-loading contaminated ballast water that had been taken on in Latin American ports. Since 1973 there have been over two hundred proven cases of cholera in the U.S. and probably many more that were not reported.

Cholera has been rare in all industrialized nations for the last one hundred years, but it is still common in certain parts of the world, including sub-Saharan Africa and the Indian subcontinent. The seventh pandemic, which began in Southeast Asia in 1961, is still ongoing today in Asia, Africa and Latin America. It shows no signs of subsiding, and will not do so without adequate sanitation. This pandemic reached Peru, South America, in 1991. The outbreak quickly increased to epidemic proportions and spread to other South American countries, Central America, and Mexico, in spite of efforts to prevent it. From January 1991 to July 1995, there were over one million reported cases of cholera in the Western hemisphere, and over 10,000 deaths, demonstrating the potential devastation of this disease. In Peru alone, over 250,000 symptomatic cases were reported. The ratio of symptomatic to non-symptomatic cholera is approximately 1:400, meaning that for every one infected person who has diarrhea, there are four hundred infected individuals who have no symptoms. At that rate, 100 million people in Peru were most likely infected with *V. cholerae*.

During the current pandemic, cholera has been brought into the United States by travelers returning from afflicted countries, or by contaminated seafood they brought back with them—over one hundred cases have been attributed to the South American strain. This strain has been isolated from the Gulf Coast, but no cases of cholera have been associated with seafood harvested from U.S. waters. The normal habitat for cholera organisms is warm coastal water, and they cannot be eradicated from these areas.

In the fall of 1993, a new strain of *V. cholerae* (O139) was implicated in an outbreak of cholera in Bangladesh and India, where it is now considered to be endemic. Only one U.S. case in a traveler returning from India has been due to this strain, which is not present in U.S. waters. *V. cholerae* O1, on the other hand, is a major cause of epidemic diarrhea in other parts of the world. It is also responsible for the ongoing global pandemic in Asia, Africa, and Latin America, which has already lasted forty years. Untreated, cholera caused by this organism is twenty-five to fifty percent fatal.

Characteristics of *Vibrio cholerae*
These bacteria are unique in their shape as short, curved rods. They are motile, facultative, and grow best at a neutral pH but also tolerate alkaline conditions. They are resistant to freezing for up to four days, but are readily killed by drying, short exposure to disinfectants including chlorine, dry heat of 243°F, steam and boiling. *Vibrio* bacteria cannot live in pure water, but will survive for up to six weeks in water containing organic matter, or in sewage for twenty-four hours.

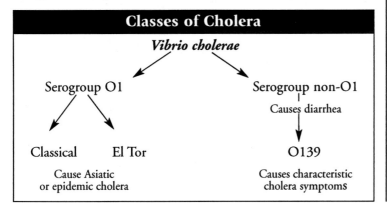

Classes of Cholera

Vibrio cholerae

Serogroup O1   Serogroup non-O1
Causes diarrhea

Classical   El Tor   O139

Cause Asiatic
or epidemic cholera

Causes characteristic
cholera symptoms

The species *Vibrio cholerae* is divided into strains (serogroups) based on differing antigenic properties. Two of these serogroups are agents of cholera in humans, O1 and O139. The O1 serogroup, members of which have been responsible for the seven pandemics and are found in U.S. coastal waters, is further divided into "classical" and "El Tor" strains. Cholera caused by the O1 strains is called "Asiatic" or "epidemic" cholera.

Non-O1 strains of *V. cholerae* typically cause gastroenteritis in humans and other primates though O139 infection produces characteristic cholera. Those with preexisting liver problems, such as cirrhosis, are more susceptible to septicemia, which is seen only rarely with non-O1 strains. The FDA advises such individuals to avoid raw and undercooked shellfish. Non-O1 strains are related to the O1 serogroup, but the disease they cause is less severe than cholera. Pathogenic and nonpathogenic strains are also found in U.S. coastal waters. When ingested in large numbers (over 1,000,000 organisms) in raw or undercooked seafood, particularly shellfish, these non-O1 strains cause diarrhea, which sometimes contains blood and mucus, abdominal cramps and fever. In twenty-five percent of victims, symptoms will also include nausea and vomiting. Symptoms begin within forty-eight hours of ingestion and may last up to a week. This type of food poisoning from raw oysters has been noted to increase in frequency in the warmer months due to seasonal turnover of ocean water, and is the basis for the anecdotal advice that raw oysters not be consumed during the months that do not contain the letter "R." There have been no major U.S. food poisoning outbreaks due to non-O1 *V. cholerae* strains, but sporadic cases do pop up along the coasts.

Another species of *Vibrio* that causes food-borne gastroenteritis is *V. parahemolyticus*, which is the leading cause of food poisoning in Japan due to its association with raw fish. It causes watery diarrhea accompanied by abdominal cramps, nausea, headache, and sometimes vomiting. The illness lasts one week or less and may result in death of older victims.

V. vulnificus is also found in seafood and seawater, and lives in the warm waters of the Gulf Coast. Its nickname, "Terror of the Deep," is an indication of its pathogenic potential—it is significantly more virulent than *V. parahemolyticus*, and

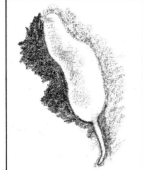

Vibrio
vulnificus

if ingested with raw or undercooked shellfish, particularly oysters and clams, it may cause a rapidly progressive septicemia that is fatal in forty to sixty percent of cases. It is one of the most invasive organisms ever described by bacteriologists. Though infection is rare among healthy people, individuals with liver dysfunction are particularly vulnerable. *V. vulnificus* also causes serious wound infections that can be acquired while swimming or cleaning shellfish or crabs. Such infections often require amputation of the affected limb.

Symptoms of Cholera

Cholera is rarely traceable to a single source, but in epidemics, it is usually due to fecal material that contaminates drinking water. Other sources include food and dirty hands or utensils. In the U.S., the Centers for Disease Control and Prevention report zero to five cholera cases annually, primarily among travelers. The risk of contracting cholera is about one in one million. It is not spread by contact with an infected person, and one million bacteria must be ingested to overwhelm the body's defenses and produce symptoms. Victims may lose as much as ten percent of their body weight in fluids, and death can occur in hours or days.

V. cholerae is not typically invasive, but it has the ability to adhere to intestinal mucosa and to produce a potent enterotoxin, simply referred to as "cholera toxin." Cholera toxin causes diarrhea by disrupting the normal osmotic balance control systems of the small intestine. The incubation period can vary from six hours to five days, depending upon the dose of organisms ingested. There is a sudden onset of headache, leg cramps, nausea and vomiting but little or no fever, followed by severe, though painless, diarrhea. The severity of infection ranges from mild or even asymptomatic, to diffuse and watery diarrhea referred to as "rice water" that can be so voluminous that the lower intestine is unable to reabsorb excreted fluids. Fluid loss, which may approach ten quarts per day without treatment, can lead to shock, toxemia and eventually death within hours in twenty-five to fifty percent of cases due to dehydration. With prompt treatment, the death rate drops to less than one percent.

Everyone is susceptible to cholera, but there is some variability in natural resistance. Normally, the acidity of the stomach is very high. This acidic environment helps to destroy *V. cholera* and other pathogens. The use of antacids has been linked to increased susceptibility to cholera due to decreased acidity of the stomach. Individuals with normally reduced gastric acid seem to be more vulnerable to cholera infection, as do the immunocompromised and the malnourished. An asymptomatic carrier state is also possible, wherein an individual harbors the organism but shows no symptoms of illness. These carriers will still shed the bacteria in their feces. Survivors of cholera will be immune to reinfection, perhaps for years.

Diagnosis and Treatment of Cholera
The primary clinical symptom that is useful in diagnosing cholera is "rice water" stools without blood. Motile *V. cholerae* bacteria can be seen in microscopic examination of patient's stool specimens, and can be cultured on differential media. *V. cholerae* can also be isolated from suspected water or food samples, but must be assayed for the production of cholera toxin since nonpathogenic strains will not produce it.

Treatment must include replacement of fluids and electrolytes. A prepackaged, dry mixture of salt, sodium bicarbonate and glucose that is reconstituted in water is used to treat cholera worldwide. This can be administered orally or, if the patient is vomiting or if diarrhea is particularly severe, intravenously. Antibiotics will effectively shorten the duration and reduce the severity of illness, but are not as important as rehydration.

Diagnosis and Treatment of Gastroenteritis
Gastroenteritis caused by non-O1 strains of *V. cholerae* can be diagnosed by finding pathogenic bacteria in the patient's stool, or in the case of septicemia, the patient's blood. *V. cholerae* gastroenteritis can also be treated with antibiotics.

Preventative Measures for Cholera
Travelers to countries where cholera is endemic, such as Latin America, Africa and Asia, should take precautions regarding drinking water. Unlike crypto, *V. cholerae* is susceptible to chemical treatments, so iodine or chlorine will destroy it, as will boiling. Other beverages that are safe include tea and coffee made with boiled water, and carbonated beverages served *without ice*. Foods must be cooked. Raw vegetables, salads, raw or undercooked fish or shellfish should not be eaten. Raw fruits may be eaten if you peel them yourself. Foods from street vendors should also be

avoided, and foods should not be brought back to the U.S. Take the advice offered by the CDC, "Boil it, cook it, peel it or forget it." These recommendations will protect against cholera and other agents of what is collectively known as "traveler's diarrhea."

U.S. and international public health authorities are working to improve surveillance and epidemiology of cholera outbreaks, and to design and implement preventative measures. The U.S. Agency for International Development is providing assistance to countries suffering from endemic cholera. Within the U.S., the Environmental Protection Agency is working with sewage and water treatment facilities to prevent contamination of water supplies, and the Food and Drug Administration tests imported and domestic shellfish for *V. chloerae* and monitors shellfish beds per the Shellfish Sanitation Program. The CDC continues to investigate outbreaks and train lab workers, and provides information to health officials and the public.

According to the CDC, the only licensed cholera vaccine in the U.S. is no longer being manufactured. This vaccine had provided only about fifty percent protection for a period of six months, and was not recommended for travelers due to this brief and incomplete immunity. Two other types of vaccine are licensed and available in other countries. Both are more effective and last longer than the discontinued version, but neither vaccine is recommended for travelers nor available in the U.S. No vaccines currently exist for strain O139, and other cholera vaccines are not effective against it. Primary infection with O1 will not confer immunity to O139.

Cholera epidemics occur in areas of poverty and poor sanitation. While rehydration can save lives, it is often difficult to deliver the necessary treatment to areas where it is needed. Unfortunately there is also increasing antibiotic resistance among *V. cholerae* strains as well.

Sabotage of Food & Water with *V. cholerae*

Military experts say *V. cholerae* is *not likely* to be dispersed as an aerosol because it is unstable in this form, does not survive well in the environment, and is easily killed with disinfectants. Therefore, the major threat appears to be contamination of water supplies and possibly food. The best recourse is to increase security measures at food plants and municipal water facilities. Cholera is not contagious, so to be effective, contamination levels would have to be high, especially in water supplies, due to the susceptibility of the organism to chlorine. Even so, early diagnosis and treatment would effectively keep death rates low.

Tularemia

History

Tularemia was first described in 1911 by George McCoy, who was studying a plague-like illness in ground squirrels following an earthquake in San Francisco that prompted an outbreak of bubonic plague. He named the agent of "squirrel plague" *Bacterium tularense* after Tulare, California, where his lab was located. The first human case was confirmed in 1914.

A few years later, Dr. Edward Francis, who studied a disease known as deer fly fever, discovered that the causative agent of his disease and that of McCoy's squirrel plague was one and the same, and that deer flies served as carriers. In 1921, Francis named the disease *tularemia*, and in 1959, he was awarded the Nobel Prize for his work on the disease. The bacterium responsible for the disease was renamed *Francisella tularensis* in his honor.

Epidemics of tularemia in the 1930s and 1940s brought to light the devastating potential of this organism, with large water-borne outbreaks in Europe and the Soviet Union, and epizootic cases (affecting many animals of one kind at the same time) in the U.S.—1939 was a record year with 2,291 cases reported. U.S. incidence remained high through the 1940s, with over 1,100 cases annually. The decline seen in the 1950s and 1960s is purportedly due to several factors, including a decrease in rabbit hunting, and perhaps failure on the part of physicians to recognize the disease. Since 1967, less than two hundred cases have been reported annually in the U.S. with a death rate of 1.4 percent. (But from 1985 to 1992, there were 1,409 cases and twenty deaths). Most cases were among those younger than ten and older than fifty, and seventy-three percent of victims were male.

Tularemia is usually spread by direct contact with the organism via infected animals, contaminated water, or bites of insect vectors, but it can also be inhaled. The largest airborne outbreak of tularemia occurred in Sweden in 1966 and 1967. The organism became airborne in dust generated by routine farming practices and infected over 600 people.

Francisella tularensis: **The Organism**

This bacterium is a small, oval-shaped bacillus that often varies in size and shape. It is aerobic (requires oxygen) and nonmotile (lacks flagella). It does not form spores, yet will survive for extended periods under cold, moist conditions. *F. tularensis* is resistant to freezing and remains viable for weeks in the soil, water, hay or straw, or animal hides and carcasses. Heat and disinfectants will readily kill this organism.

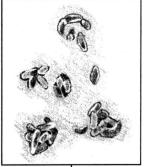

Francisella tularensis

F. tularensis occurs as two different variants, *F. tularensis tularensis* and *F. tularensis palearctica*. The former, also known as type A, is most prevalent in the U.S., primarily in Missouri, Illinois, Virginia, Tennessee, Kansas, Oklahoma, South Dakota, Arkansas and Texas. The tick serves as its primary reservoir although the bacterium has been isolated from fifty-five other arthropods and over one hundred other animals. Type A is highly virulent for rabbits and humans, and has been recovered from several types of wild rodents, ticks and deer, and from domestic animals including sheep, cattle, horses, swine, and even cats and dogs. The second type, which can be recovered from water, mosquitoes, water rats and other aquatic mammals, is more common outside the U.S. and is relatively avirulent (not very likely to cause illness) for rabbits and humans. Water-borne, air-borne and arthropod-carried tularemia outbreaks involving hundreds of cases have been reported in Eurasia, but incidence worldwide is not really known since the disease is most likely underdiagnosed and underreported.

Tularemia: **The Disease**

Tularemia is a zoonotic disease (one that can be transmitted from animals to humans) that is endemic in North America, Europe, the Middle East, Russia and Japan, with the greatest number of cases being reported in northern and central Europe, especially Scandanavia, and the former Soviet Union. It is rare in the United Kingdom, Africa, Central and South America. U.S. tularemia cases are sporadic and often occur in clusters due to contaminated drinking water, or exposures among laboratory workers—numerous cases have been due to the processing of infectious samples or while doing research. Inhalational tularemia has been acquired in the lab by simply examining an open *F. tularensis* culture dish. All U.S. states except Hawaii have reported cases of tularemia.

Naturally occurring tularemia is almost exclusively a rural illness, transmitted to humans primarily through handling of contaminated rabbit carcasses, or through deer fly or tick bites. Hunters contract the disease through breaks in the skin or via the mucous membranes of the eye when they clean rabbit or other rodent carcasses, the source of the organism being the blood, tissue or body fluids of infected animals. Tularemia has also been called "rabbit fever" and "deer fly fever" because bites of deer flies, mosquitoes or ticks can also transfer the bacteria. Cases have also been linked to muskrat handling, lawn mowing and brush cutting. (Fecal material of ticks is most likely the source of the bacterium as it is not found in tick saliva.)

Less commonly, *F. tularensis* can cause an inhalational disease from contaminated dust. It is one of the most pathogenic bacteria known—ten to fifty organisms inhaled or injected under the skin have been shown to initiate tularemia.

Intestinal illness results from ingestion of *F. tularensis* organisms in contaminated food or water. Everyone is susceptible to tularemia, but those engaged in hunting, trapping, butchering, and farming are more likely to be exposed, with adult men and laboratory workers identified as high-risk.

F. tularensis is an intracellular bacterium that multiplies within macrophages, spreads to regional lymph nodes and may become systemic early in the course of infection as bacteria target the lungs and surrounding tissues, kidneys, liver and spleen. Tularemia manifests in several forms. The most common, *ulceroglandular,* results from a tick, mosquito or deer fly bite, or can be contracted through even the tiniest break in the skin. It accounts for seventy to eighty-five percent of tularemia cases. In sixty percent of ulceroglandular cases, an ulcer forms at the site of bacterial entry. These ulcers can be up to one inch wide, with raw centers and ridged, crater-like edges. Regional lymph nodes become swollen eighty-five percent of the time, and the patient experiences fever, chills, sweating, headaches, muscle pain and coughing, and may have vomiting, sore throat, chest, joint, back or abdominal pain, or a stiff neck. Very often, if the vector was an infected animal carcass, upper extremities will be affected, whereas arthropod bites typically affect lower extremities. Untreated, ulceroglandular tularemia has a four percent mortality rate. Five to twelve percent of tularemia cases are of the *glandular* variety, wherein lymph nodes swell and the patient will have a fever, but there is no apparent skin lesion.

Oculoglandular tularemia is contracted through the mucous membrane of the eye and results in severe conjunctivitis (an inflammation of the eye tissue) accompanied by swollen lymph nodes. This form represents one to two percent of all tularemia cases. *Oropharyngeal* tularemia is usually caused by ingesting, and sometimes by inhaling, the organism. With this type of tularemia, a lesion forms in the oropharynx (back of the throat) and causes bilateral tonsillitis or a severe sore throat that resembles strep throat, and the patient has debilitating headaches. Lymph nodes in the neck swell one to two weeks following the initial infection and may persist for some time.

Typhoidal tularemia has the potential to be severe and sometimes fatal. It is acquired from inhalation of *F. tularensis,* and represents seven to fourteen percent of cases. This systemic form of the disease has a rapid onset of fever, chills, and headache, and is the only type of tularemia with gastrointestinal symptoms of abdominal pain, vomiting and diarrhea. It is also the only form lacking the symptom of swollen lymph nodes, and there is no sign, such as a lesion, to indicate the site of inoculation. Typhoidal tularemia has a relatively high mortality rate of thirty-five percent, because eighty percent of the time it progresses rapidly to the final type of tularemia, *pneumonic.*

Eight to thirteen percent of tularemia is pneumonic. This type results as a complication of the other types, particularly typhoidal, and affects the lower respiratory tract. Lymph nodes in the lungs become swollen, and the primary symptom is a productive or nonproductive cough, sometimes accompanied by shortness of breath and/or chest pain. Infrequently, pneumonic tularemia can lead to complications such as inflammations of the pericardium, gastrointestinal tract, appendix, or meninges. Severe pneumonic tularemia may result in respiratory failure and death.

With treatment, tularemia in general has a mortality rate of one to two-and-a-half percent. In all types of tularemia, nodes may remain painful and swollen for long periods—up to three years—and eventually may become necrotic and drain. Incubation is usually three to five days but varies widely and is dose dependent.

All ages are susceptible to tularemia, but it is most prevalent in those ages five to nine and over seventy-five. In all age groups, males are more susceptible. Exposure and recovery confers lifelong immunity.

Diagnosis and Treatment of Tularemia

Clinical symptoms are not sufficient to diagnose tularemia. Symptoms of typhoidal tularemia will resemble those caused by *Salmonella typhi*, rickettsia or malaria, and pneumonic tularemia symptoms will mimic those due to plague or inhaled staphylococcal enterotoxin B (SEB), although tularemia can cause a nonproductive cough in association with pneumonia. Lab tests are not very helpful, but about fifty percent of patients' chest X-rays, most commonly with typhoidal tularemia, will show evidence of pneumonia or swollen lymph nodes in the chest. Sepsis resulting from a tularemia infection has the potential to be severe and sometimes fatal. Patients typically will appear toxic and may be confused, perhaps even becoming comatose. Without prompt intervention, death will result from septic shock, acute respiratory distress and organ failure.

F. tularensis may be present in patient's sputum, lymph node material, gastric washings, wounds, throat or eye secretions and occasionally, blood. It is difficult to culture in the lab because it has peculiar growth requirements. It is highly contagious, so extra handling precautions must be taken, including the use of a BL-3 (biohazard level 3) containment hood to prevent aerosolization of the organism. Even if it grows on prepared media, its growth is often overshadowed by that of normal human flora (bacterial populations), but growth of *F. tularensis* in culture is the definitive means of confirming the diagnosis of tularemia.

The same patient samples can be assayed using PCR or fluorescent antibody analysis, a rapid diagnostic procedure that is available in some reference labs. Serology can be done retrospectively to confirm tularemia, about two weeks after the initial infection, but most diagnoses are made with the use of bacterial agglutination or ELISA tests, which were made available for *F. tularensis* just recently.

Before antibiotics became available, type A tularemia had an overall mortality of five to fifteen percent, which climbed as high as sixty percent in severe pneumonic and systemic forms. With antibiotic treatment, those rates have decreased to one to two-and-a-half percent for all types. Certain types of antibiotics can also be used to prevent tularemia. Treatment of the disease should include management of secretions and lesions, but *F. tularensis* is typically not spread person-to-person via respiratory droplets so patient isolation is not necessary.

Preventing Tularemia
As of June 2001, a live, attenuated (unable to cause illness) vaccine for tularemia was still being reviewed by the FDA. It had proven effective in preventing lab-acquired tularemia, and was recommended for those at risk. Volunteer studies, however, have shown that this vaccine is not completely protective against all aerosol challenges. The degree of protection may be dependent upon the dose of *F. tularensis*—extremely high doses could be overpowering. Common disinfectants and heat, 145°F for 30 minutes, will kill *F. tularensis*. Hands and clothing should be washed with soap.

Tularemia as a Biological Weapon
Japanese germ warfare research units operating in Manchuria studied *F. tularensis* between 1932 and 1945. During World War II, tens of thousands of Soviet and German soldiers, as well as civilians in Eastern Europe, were affected with tularemia that may have been the result of intentional dissemination of *F. tularensis*.

F. tularensis was weaponized by the United States in the 1950s and 1960s. It was produced at the Pine Bluff Arsenal in Arkansas, and twenty grams of *F. tularensis* were among the bioweapons retained by the Central Intelligence Agency after President Nixon's 1969 ban. During the same period, the U.S. also worked on developing antibiotics and vaccines for tularemia. *F. tularensis* still remains on the list of potential biological weapons that could be used against the U.S. since other countries supposedly have weaponized it as well.

The potential ramifications of an intentional release of *F. tularensis* was summarized in 1969 by the World Health Organization, who estimated that about 110 pounds disseminated over a metropolitan area with five million people would leave 250,000 incapacitated and 19,000 dead. Tularemia would last several weeks, with relapses for weeks or months. The economic impact of such an attack, according to the CDC, would be $5.4 billion for every 100,000 people exposed.

The Russians tested tularemia on Renaissance Island (Vozrozhdeniye Island in the Aral Sea) in the 1980s, along with smallpox, Q fever, brucellosis, glanders and plague. Their research continued into the early 1990s and included development of strains that were resistant to antibiotics and vaccines. As recently as 1999, the U.S. still produced 1.6 metric tons of *F. tularensis*, and the Russians, 1,500 metric tons, annually.

According to a recovered invoice, in 1984, the Ranjeeshee cult that had contaminated salad bars in Oregon with *Salmonella* had ordered other microbes from the American Type Culture Collection (ATCC) as well, including *F. tularensis*. It is estimated that within thirty-six hours, a terrorist could grow enough bacteria to produce about five quarts of material. A garden sprayer could be used to deliver enough agent through an air-intake system to infect half of the people in a building the size of the former World Trade Center.

Equally as frightening, according to the Working Group on Civilian Biodefense, *F. tularensis* has been engineered to be resistant to the antibiotics chloramophenicol and tetracycline, and virulent streptomycin-resistant strains have been studied as potential agents of biowarfare by both the United States and the Soviet Union. The Working Group also states that although *F. tularensis* virulence factors are poorly understood at this time, it is possible that they could be enhanced through laboratory manipulation.

F. tularensis can be stabilized and produced in wet or dry form. Delivery would presumably be as an aerosol containing viable organisms. Within three to five days, there would be an outbreak of inhalational tularemia, though large amounts of this aerosol may shorten the typical incubation time. Studies have shown that inhalational tularemia is a systemic illness though different from typhoidal tularemia, with prominent signs of respiratory disease. Inhaled tularemia is capable of incapacitating some individuals within the first one or two days after exposure, with continued impairment for several days even after antibiotics are started. Untreated, symptoms would persist for weeks or months and grow continually worse. Other types of tularemia, such as oculoglandular, ulceroglandular, glandular or oropharyngeal, may also result from aerosol exposure. Any form could develop into severe pneumonia (pneumonic tularemia), sepsis or meningitis, and have a mortality rate of over five percent.

Animal infections may also occur, especially among rabbits, so the skinning and eating of rabbits should be avoided in the event of an intentional release. Water and grain would also be contaminated and would need to be boiled or cooked prior to consumption. Organisms would remain viable in the soil for only a short time, and would probably not present a significant hazard. An aerosol attack of significant proportions may, however, leave high enough numbers of bacteria in the environment that they could present a threat through secondary exposure.

Francisella tularensis as a Food or Water Contaminant

The Working Group has determined that an aerosol release of *F. tularensis* would have the greatest adverse effects, but aerosols would probably have a short half-life in the environment since the organism is susceptible to drying and solar radiation.

Humans can contract tularemia through ingestion of contaminated food or water, but contamination of drinking water would not be effective as a means of dispersal because this bacterium is killed by chlorine at concentrations commonly used in municipal water supplies. A food vehicle may be a possible alternative, resulting in *oropharyngeal tularemia*. Though it could lead to more serious conditions, early diagnosis and treatment would prevent complications and keep mortality very low.

In order to better prepare for an attack with *F. tularensis*, the Working Group believes that better rapid tests are needed for use in the event of mass exposures, as are more methods to rapidly identify organisms that differ genetically from native strains of *F. tularensis*—strains that may possess enhanced virulence traits, or resistance to antibiotics or environmental factors. Research is also needed to develop rapid methods for detecting the organism in the environment, and possibly in food and water samples.

Water-borne Pathogens

Disease	Symptoms	Transmission	Prevention / Treatment	Food / Water Sabotage
Cryptosporidiosis *(Cryptosporidium parvum)* (parasite)	Watery diarrhea, stomach cramps, low fever, possibly vomiting & weight loss; can also be asymptomatic	Contaminated water or raw fruits & vegetables washed in contaminated water; contact with infected animals or people; can also be inhaled (rare)	Avoid contaminated water, wash hands / replace fluids & electrolytes, antidiarrheals per doctor's advice	Possibly fresh fruits & vegetables & other foods that are not heated / Possible—has occurred naturally
Giardiasis *(Giardia lamblia)* (parasite)	Same as above	Water or food contaminated with fecal material	Avoid contaminated food & water / Replace fluids & electrolytes, may use anti-parasitic drugs	Same as above
Cholera *(Vibrio cholerae)* (bacterium)	Profuse, watery (rice water) diarrhea	Water contaminated with human feces; raw or undercooked seafood, especially shellfish	Avoid contaminated water & raw or undercooked seafood / Replace fluids & electrolytes; May use antibiotics	Possible / Possible but need high level since it is killed by chlorine
Gastroenteritis *(Vibrio parahemo-lyticus)* (bacterium)	Watery diarrhea, cramps, nausea, headache, sometimes vomiting	Water contaminated with human feces; raw/under-cooked seafood	Same as above	Same as above
Ulceroglandular tularemia—most common type *(Francisella tularensis)* (bacterium)	Skin ulcer, swollen nodes, fever, chills, sweating, headache, myalgias	Handling contaminated animal carcasses, bite of deer fly or tick	Avoid infected carcasses; wash hands & clothes / Antibiotics	Possible / Possible but need high levels since it is killed by chlorine (aerosol release also possible)

CHAPTER EIGHT

E. coli O157:H7 and *Shigella*

These bacterial species are not on the "most wanted" list of potential bioterrorist's weapons. They have not been the subjects of discussions and reports by the Working Group on Civilian Biodefense, nor are they

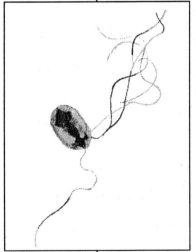

E. coli
O157:H7

talked about in the books on terrorism that have emerged since September 11, 2001. Yet to the microbiologist, these organisms are among the most feared water-borne and food-borne pathogens known. In the United States, a relatively benign pathogen, *Campylobacter* (discussed in Chapter Two) is the primary cause of food-borne infection, but much more ominous bacteria take the second, third and fourth spots: *Salmonella* (discussed in Chapter Two), *E. coli* O157:H7, and *Shigella*, respectively.

While *E. coli* O157:H7 has only recently been identified as a killer, primarily of children, *Shigella dysenteriae* has been recognized for decades as the causative agent of severe, life-threatening dysentery. These microbes have two deadly traits in common—the ability to invade mammalian cells and to produce the same deadly enterotoxin.

As they normally occur in foods and in water, these pathogens can be dealt with through proper cooking, adequate sewage systems and water treatments, and by preventing cross contamination in the kitchen. But as weapons of sabotage, they could produce devastating affects.

Historical Perspectives

E. coli O157:H7

Of all the long and complex names in the science of microbiology, this genus name, *Escherichia* (pronounced esh-er-ik-ee-ah) is probably one of the most difficult to pronounce. Therefore, the now-famous organism is commonly referred to as simply *E. coli*. Its name comes from an Austrian doctor, Theodore Escherich, who originally isolated this bacterium in 1885, though he named it *Bacterium coli commune* because it was commonly found in the intestinal tract of man and other animals.

January 1993 will forever be associated with a newly recognized pathogen: *E. coli* O157:H7. That was the month of the Jack-in-the-Box outbreak involving ninety-three restaurants, the outbreak that affected over 400 people in four states (Washington, Idaho, California and Nevada), leaving fifty victims with hemolytic uremic syndrome (HUS) and four young children dead. It was the one that made headlines, even though it was the sixteenth outbreak of O157:H7 since its discovery in 1982, and the sixth outbreak associated with undercooked ground beef. This was also the outbreak that resulted in an important change in cooking parameters for ground beef. Prior to this incident, the Food and Drug Administration had specified that hamburgers be cooked to a minimum internal temperature of 140°F. This was subsequently changed to 160°F. Since the Jack-in-the-Box episode, there have been at least one hundred more O157:H7 outbreaks in the U.S., and the United States Department of Agriculture believes that the numbers will continue to rise.

Well before the Jack-in-the-Box fiasco, an outbreak in Alberta, Canada, in 1977 that was traced to hamburgers from McDonald's, led to an investigation by Canadian medical researchers. Dr. Mohammed Karmali connected the severe diarrheal illness experienced by victims who had eaten hamburgers to contamination with *E. coli*. But the serotype (O157:H7) was not identified until 1982. This particular *E. coli* produces an enterotoxin that was also identified in 1977 by another Canadian, food microbiologist Dr. J. Knowalchuk. He found that particular strains of *E. coli* were capable of producing a toxin that was lethal to a certain line of tissue culture cells, African green monkey kidney cells called "vero cells." The toxin was therefore dubbed "verotoxin." Meanwhile, in the U.S., scientists discovered this toxin to be very similar to that produced by *Shigella* bacteria, so they named it "Shiga-like toxin," or SLT. To date, over one hundred *E. coli* serotypes have been found to produce the deadly toxin.

The nineties were busy years for O157:H7 as well. Outbreaks in Missouri and Utah, both due to contaminated drinking water, affected over 250 people and left eight dead. Municipal water in Cabool, Missouri, became contaminated when pipelines froze and burst. In July of 1993, a similar situation occurred in New York City when O157:H7 found its way into the water supply and managed to survive chlorine treatment. In Oregon, people became ill after swimming in fecally-contaminated lake water.

Over ten thousand people in Japan, most of whom were school children in the Sakai district, became ill in July of 1996 after consuming sushi, radish sprouts and raw liver, at least one, if not all, of which were contaminated with *E. coli* O157:H7. Over 500 children were hospitalized with a serious illness, fifty of them were gravely ill, and eleven eventually died from complications of infection with O157:H7. The Sakai School Lunch Program Association Director, who felt responsible for the outbreak, committed suicide.

In late 1999, Supreme Beef Processors, Inc. of Dallas voluntarily recalled approximately 180,000 pounds of ground beef after USDA tests discovered a sample that was positive for the presence of *E. coli* O157:H7. The beef was produced on December 20, but was not tested until December 25, so the product had already been distributed throughout Texas, Oklahoma, Arkansas, Louisiana, Tennessee, Mississippi, Florida, and New Mexico. Luckily there were no outbreaks associated with the implicated beef.

Municipal water was once again the vehicle in May 2000 when over 2,000 residents of Walkerton, Ontario, became ill, and twenty died. It is supposed that, similar to the *Cryptosporidium* outbreak in Milwaukee, heavy rainfalls caused excess runoff from cattle yards and introduced O157:H7 to the drinking water supply.

These are just a few of the outbreaks reported in recent years. *E. coli* O157:H7 is not only on the list of emerging pathogens, its repeated and deadly appearance worldwide has placed it very near the top. Its role in severe food-borne disease has increased steadily since 1982, and the trend continues. The Centers for Disease Control and Prevention estimate there are 73,000 cases, 2,100 hospitalizations and 250 to 500 deaths each year in the United States due to O157:H7 poisoning, which is (understandably) a reportable illness in this country. Worldwide, O157:H7 infection has occurred in thirty countries on six continents.

Shigella species

While *E. coli* O157:H7 is a recently discovered food- and water-borne pathogen, *Shigella* has been around for a while. It was first isolated in 1898, and in 1917, it was discovered to be the agent of an epidemic of diarrhea in a World War I Eastern European prison camp. Another outbreak occurred in a psychiatric hospital in Wales in 1938.

The species responsible for these outbreaks was originally named *Bacillus dysentariae*, but in 1950, it was renamed *Shigella dysentariae* after the Japanese bacteriologist Shiga, who first isolated the organism in 1898, and associated it with dysentery in humans. Dysentery is more severe than diarrhea, resulting in loss of copious amounts of fluids, as well as blood and mucous. The illness *S. dysentariae* causes is known as "shigellosis" or "bacillary dysentery." Prior to the availability of antibiotic treatments, shigellosis killed up to twenty percent of its young victims, and fifteen percent of adult victims. Less virulent species of *Shigella* could cause death ten percent of the time. Unfortunately, in the early 1960s, *S. dysentariae* became the first bacterial agent of diarrhea to develop resistance to penicillin.

Resistance to penicillin and other antibiotics is another characteristic that is often carried on plasmids (non-chromosomal, circular pieces of DNA). Antibiotic resistance is considered to be a virulence trait, similar to the ability to invade or produce a toxin. As discussed in *The Coming Plague* by Laurie Garrett (1994), an epidemic caused by a strain of *Shigella* that was resistant to a wide variety of antibiotics began in September of 1983 among members of the Hopi Indian tribe in Arizona. Over a three-year period, one woman of the tribe had taken many different types of antibiotics to fight recurring urinary tract infections. After she became infected with *Shigella*, the particular strain, in the presence of all the antibiotics in the woman's system, developed multiple resistances.

The organism was then contracted by other members of the tribe, and by 1986, was responsible for seven percent of the cases of shigellosis nationwide. Within four years, this "super shigella" had reached Canada.

Prior to 1945, *Shigella* was transmitted primarily through water contaminated with human waste, but since then, infected food handlers are mostly responsible. As with so many food-borne illnesses, a large portion of *Shigella* infections go unrecognized, making the actual number much larger than those reported. Each year in the U.S., 18,000 cases of shigellosis are reported, but the Centers for Disease Control and Prevention estimate that there are actually about 300,000. *Shigella* infections run rampant in developing countries with poor sanitation, where they occur on a steady basis in most communities.

Organismal Characteristics

E. coli

Of all the bacterial species known, *E. coli* has gotten the most publicity. This small bacilli was the first bacterium to be genetically characterized. It is also easy to grow in the lab, and has been used extensively in cloning experiments. Scientists have successfully inserted various genes into *E. coli* bacteria, which are now used to produce beneficial products like insulin and rennin, a milk coagulant used in the manufacture of cheese.

Nonpathogenic species of *E. coli* are normal, symbiotic inhabitants of the gastrointestinal tract of healthy humans and other animals, where they perform essential functions such as synthesis and absorption of certain vitamins, the breakdown of cellulose, and competitive exclusion of pathogenic bacteria. They are among the "good" bacteria, and mankind would not fare very well without them. Since they are commonly found in fecal material, *E. coli* and similar types of bacteria are used as *indicators* of fecal contamination, meaning that their presence in food or water is an indication that other pathogenic intestinal bacteria, such as *Shigella* or *Salmonella*, may also be present. For this reason, indicator organism counts are used to monitor the safety of lake water for swimming, and to determine the efficiency of sanitation measures used in food processing environments.

While the majority of *E. coli* bacteria are beneficial, some have resorted to a life of pathogenicity. There are four different categories of pathogenic *E. coli*: enteropathogenic *E. coli* (EPEC), which is the primary cause of traveler's and infant diarrhea; enterotoxigenic *E. coli* (ETEC), which produces a diarrhea-causing toxin; enteroinvasive *E. coli* (EIEC), which can severely damage intestinal tissue; and enterohemorrhagic *E. coli* (EHEC), the agent of bloody diarrhea and the category to which O157:H7 belongs.

The serologic designation (O157:H7) is based on certain antigens present on this particular type of *E. coli* bacterial cell. Antigens are specific markers that interact with particular antibodies, and are therefore useful in categorizing bacteria and in identifying agents of disease. The "O" antigens, also referred to as somatic (body) antigens, are present on the bacterial cell wall. "H" antigens are present on flagella, the whip-like structures that enable certain bacteria to move. These antigens are given numerical designations, so O157:H7 means that this particular serotype of *E. coli* has somatic (O) antigen number 157, and flagellar (H) antigen number 7. This same methodology is important in identifying other bacteria as well, particularly *Salmonella*.

Characteristics of *E. coli* O157:H7

This bacterium is the most deadly of all *E. coli*. A colleague and food safety consultant from Florida, Dr. Steve Goodfellow, refers to this organism as "*Shigella* in an *E. coli* package." *E. coli* O157:H7 is indeed a renegade, quite unlike the nonpathogenic varieties of *E. coli*, and more virulent than the other pathogenic forms. It is found in the gastrointestinal tract of healthy cattle that ordinarily show no symptoms of illness, but it is thought that animals subjected to stress, such as those en route to market, may have an increased risk for infection. O157:H7 has also been isolated from cattle with diarrhea, and from healthy pigs and chickens. It is very elusive and difficult to detect in host animals where its existence is transient, but it has been found in two to four percent of ground beef samples, over one percent of pork and poultry samples, and two percent of lamb samples.

Essentially, this organism is indeed a *Shigella* in an *E. coli* suit, thanks to a process called *conjugation*—basically bacterial sex—in which bacteria exchange genetic information and therefore traits. It is uncommon for chromosomal information to be exchanged through conjugation, but the transfer of plasmids occurs frequently. *E. coli* O157:H7 produces an extremely potent toxin that is virtually identical to that produced by certain strains of

Shigella dystenariae. The genes that code for the production of this toxin are found on a plasmid. It is presumed that somewhere along the line, an enteropathogenic strain of *E. coli* picked up this virulence plasmid from a *S. dysentariae* bacterium. This "vero toxin" or "Shiga-like toxin" (SLT) is produced by O157:H7 and 150 other *E. coli* serotypes, which are sometimes referred to as "VTEC" strains: **Vero Toxin E. coli.** It destroys mammalian cells, primarily of the intestine and kidney, by interfering with protein metabolism. *E. coli* O157:H7 also has the ability to adhere to and invade intestinal cells, another trait it shares with *Shigella.* It is interesting to note that enteroinvasive *E. coli* (EIEC) are also strikingly similar to *Shigella flexneri* (discussed in the following section) in the manner in which they invade mammalian cells and create illness. This is most likely due to the transfer of the virulence plasmid from *S. flexneri* to *E. coli,* since the plasmids encoding invasion in these two organisms are identical.

Unfortunately, *E. coli* O157:H7 survives both refrigeration and freezing, and contaminated foods may show no evidence of spoilage. On the other hand, it *can* be killed by heat, so thorough cooking is the only recourse to prevent food-borne infection with this deadly pathogen. Most outbreaks of this organism are due to foods that have been contaminated after cooking, or foods that are insufficiently heated or not heated at all. Even though routine chlorination of municipal water supplies should kill O157:H7, outbreaks have occurred wherein the bacterial load was so high that it was able to overpower the chlorine. Chlorine is an effective sanitizer because it is capable of breaking down protein, a major component in the cell walls of bacteria and in the outer coat of viruses. However, after a chlorine molecule attacks a bacterium or virus, it becomes inactivated. If a water supply has a particularly high load, therefore, the chlorine may be entirely used up before all organisms can be destroyed.

Characteristics of *Shigella* species

There are four species of *Shigella,* all of which are found only in the intestinal tract of man and other primates. *S. dysentariae* and *S. flexneri* have been mentioned; the other two species are *S. boydii* and *S. sonnei.* The four differ in virulence and distribution pattern. *S. dysentariae* is the most virulent, and is commonly found in developing countries along with *S. boydii. S. sonnei* is the least virulent and found most frequently in developed countries, and accounts for about sixty-six percent of shigellosis in the U.S., with *S. flexneri* being responsible for the rest.

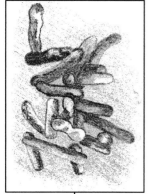

Shigella

Shigella bacteria are very closely related to *E. coli*, and are in the same family, the *Enterobacteriaceae*. Shigellae are not very resilient, and will survive for only a short time outside a host; for example, they will remain viable in stool samples for only twenty-four to thirty-six hours. *Shigella* survives refrigeration and freezing, but can be destroyed by heat. Food-borne illness by these bacteria is often due to contamination by an infected food handler.

Disease Manifestations

E. coli O157:H7—Hemorrhagic Colitis

The occurrence of this bacterium in beef cattle has led to contamination of hamburger, which is the most important food vehicle for the disease known as *hemorrhagic colitis*. All muscle tissue (meat) is sterile when an animal enters a food production line. However, during the processing operation, muscle tissue becomes contaminated with bacteria from hides, hooves, and intestines. This contamination does not pose a threat with items like steaks or roasts because the bacteria are only on the outside of these meats and will be killed in the cooking process. In the case of ground beef, however, the grinding of the muscle tissue does an excellent job of thoroughly mixing bacteria throughout the product. With the advent of O157:H7 and other pathogenic *E. coli*, the consumption of "cannibal sandwiches" (raw ground beef, raw vegetables and onions) and rare hamburgers is akin to a game of Russian roulette. While you can still get away with eating a rare steak, to be assured of bacteriological safety, hamburger should be well done throughout.

Other foods responsible for outbreaks of *E. coli* O157:H7 have included unpasteurized apple cider, raw milk, lettuce, cheese curds, game meat, alfalfa sprouts, and even dry-cured salami, proving that the organism is also resistant to acid, the fermentation process, and drying. Apple cider is usually made from windfall apples that cannot be sold as produce because they are physically bruised and damaged. On the ground, these apples become contaminated with *E. coli* and other bacteria from the soil. If the cider is not pasteurized, these bacteria survive. Most cider and juice found on grocery store shelves is pasteurized, as the label will indicate. However, products purchased at orchards or farmers' markets may not be, so beware.

Other produce, such as alfalfa sprouts, can be contaminated through the use of manure fertilizers or by flies with fecal material on their feet. Studies indicate that billions of *E. coli* and other bacteria may be present in each gram of alfalfa or bean sprouts, due to fecal contamination that can occur during handling, processing, shipment or storage. Imported fruits and vegetables are particularly vulnerable to contamination because other countries may not have such stringent regulations regarding hygiene and agricultural practices as the United States.

Contamination can also occur in the kitchen. One U.S. outbreak of O157:H7 was due to cross-contamination that occurred in a restaurant when a meat cutter assisted a waitress with a large crock of salad dressing. The crock became contaminated with the bacterium from the meat cutter's hands and subsequently contaminated other foods on the salad bar. In July of 2000, forty-two people in the Milwaukee, Wisconsin, area became ill from watermelon served at a local restaurant. The source of *E. coli* O157:H7 bacteria was traced to raw beef—another case of cross-contamination due to careless handling of foods. Of the forty-two victims, twenty-four were children. It is suspected that this deadly organism may even be a problem in petting zoos, where infection would come directly from the animal rather than food or water. It seems that *E. coli* O157:H7 is truly a formidable enemy of the young.

Hemorrhagic colitis is the primary illness caused by ingestion of as few as ten *E. coli* O157:H7 bacterial cells; it is possible that even just one cell could cause illness. This infection produces severe abdominal cramps and diarrhea that is initially watery, but sometimes becomes grossly bloody to the point that it consists of blood without fecal material. This is due to the action of the toxin produced by the organism, which damages intestinal epithelial cells enough to cause bleeding and to allow toxin to enter the bloodstream. The toxin also destroys platelets, tiny cell fragments that aid in clot formation at wound sites. Occasionally the patient will vomit as well, but fever will be low or absent. In otherwise healthy adults, symptoms will last about a week and resolve without medical intervention. Up to fifteen percent of hemorrhagic colitis victims will develop life-threatening complications from this infection.

Complications

E. coli O157:H7 kills primarily young children due to the development of hemolytic uremic syndrome (HUS), which is the leading cause of renal failure among American and Canadian children, with a death rate of five to ten percent, and twenty percent in complicated cases. It is estimated that this illness is responsible for as many as 8,000 cases and 500 deaths, primarily among children, each year in the U.S. Infection by *E. coli* O157:H7 probably causes eighty-five percent of HUS cases.

HUS was first described in 1977 by Swiss physician Dr. C. Gasser. In hemolytic uremic syndrome, red blood cells are destroyed at a rate so fast that the bone marrow, where new red blood cells are manufactured, cannot keep up and the patient becomes anemic. Hemoglobin is released from the ruptured red blood cells. This hemoglobin, as well as the shredded remains of the red blood cells themselves, accumulates in the bloodstream, choking tiny arteries and capillaries and preventing blood from reaching the organs of the body, thereby shutting them down, one by one.

The kidneys attempt to filter debris from the blood, but are quickly overwhelmed. The patient's urine becomes brown due to the hemoglobin, and the kidneys begin to shut down. HUS patients often require dialysis, blood transfusions and other supportive care such as respirators or intravenous fluids. About thirty-three percent of HUS victims will have permanent kidney damage. Adults may have seizures or stokes due to blood clots in the brain if the condition is not diagnosed in time, and other organs may be permanently damaged. In severe cases, the patient has an almost seventy-five percent chance of suffering major physical impairments such as the continued need for dialysis, or even a kidney transplant or a colostomy. About eight percent will suffer other lifelong effects such as high blood pressure, blindness, diabetes or even paralysis. The death rate of HUS is three to five percent. Susceptible adults, particularly the elderly, may develop a condition known as thrombotic thrombocytopenia purpura, or TTP, which is essentially HUS along with a fever and neurological symptoms. In the elderly, mortality due to TTP can reach fifty percent.

Disease Manifestations of Shigellosis
(Bacillary Dysentery)

Shigellae are commonly found in water that has been polluted with human feces. Therefore, infection with these bacteria has important ramifications in developing countries where *S. dysentariae* is a major cause of infant deaths due to diarrhea, with a mortality rate of up to twenty-five percent. *Shigella* infection is a problem in developed countries as well, primarily where crowded conditions prevail—schools, retirement homes, institutions, and even cruise ships. In the U.S., it is responsible for less than ten percent of food poisoning outbreaks, most of which occur in the summer. Anyone can get shigellosis, but infants, the elderly, and the immunocompromised are especially vulnerable.

All four species of *Shigella* have been implicated in food-borne outbreaks, but *S. sonnei* most frequently. The other three species are more commonly associated with water-borne outbreaks, and *S. flexneri* has been shown to be sexually transmitted as well. Each of the four species is easily transmitted person-to-person by the fecal-oral route.

Since most cases of *Shigella* food poisoning in developed countries are due to the less virulent *S. sonnei*, it is not considered to be a major disease threat in these areas of the world, though it can still cause illness. It is usually transmitted by infected food handlers, so any food has the potential to be contaminated as long as it is not heated prior to being consumed. Common vehicles include raw vegetables, all types of salads, dairy products, poultry and sandwiches. One outbreak, due to shredded, bagged lettuce from a single distributor, caused 347 cases of gastroenteritis in patrons of several restaurants. A 1988 airline outbreak that resulted in over 1,900 cases of shigellosis in twenty-four states and four countries was traced back to sandwiches made by a single food handler. Vegetables and fruits are also prone to contamination by flies that can spread the organism from fecal material and onto foods. *S. flexneri*-contaminated onions from Mexico were responsible for one U.S. outbreak of shigellosis. Contaminated foods will most likely look and smell normal.

Shigellae behave in a similar fashion to enteroinvasive *E. coli*. In the course of disease, *Shigella* organisms are ingested—as few as ten will establish infection—and attach to the epithelial cells of the large intestine. The organisms then invade and multiply within these cells, a process that results in bloody diarrhea (dysentery) as the surface layer of bowel becomes inflamed and ulcerated. The incubation period varies from one to seven days. In developed countries, the disease is self-limiting and usually runs its course within one or two weeks. Some cases may be severe, however, especially those among children and the elderly, and may require hospitalization. In children younger than two, the symptoms may include a high fever with the possible complication of hemolytic uremic syndrome. With especially virulent strains, the mortality rate may reach fifteen percent.

Complications of *Shigella* Infections

Reactive arthritis may develop in two to three percent of victims following a bout of shigellosis. Even though there are no *Shigella* organisms left in the body, they can leave their antigens behind in the patient's joints, creating inflammation and pain. Other complications include eye irritation and painful urination combined with painful joints, a condition known as *Reiter's syndrome*, which can last months or years, and may lead to chronic arthritis. However, Reiter's syndrome only develops in those individuals who are genetically predisposed to it.

An important factor in spread of shigellosis is that viable organisms will continue to be shed in the stool of recovering individuals for as long as two weeks after diarrhea is gone. Therefore, extra care must be taken so the illness is not transmitted to others. This is especially important if the patient works in the food service industry, daycare, hospital, or other type of institution.

Diagnosis and Treatment

E. coli O157:H7

Bloody diarrhea can be caused by only a few food-borne or water-borne pathogens, including *E. coli* O157:H7. The culprit can be found and identified in stool samples and incriminated food items. O157:H7 differs from other *E. coli* in that it is not capable of using the sugar sorbitol as an energy source, nor is it capable of growth at 111ºF to 113ºF (44-45ºC). These characteristics can be used to identify it in the laboratory. A special selective and differential medium, *hemorrhagic colitis aga*r, is available to determine if *E. coli* O157:H7 is present in foods.

Rapid assays are available for O157:H7, but are not currently used where they are most needed: beef production lines. Beef inspection is markedly antiquated, the quality routinely monitored only by odor and appearance, so bacterial contamination cannot be detected. Fecal material readily contaminates carcasses with *E. coli* and possibly other intestinal pathogens. Spot checks of meat samples may reveal a problem, but the chances of finding O157:H7 are akin to finding a needle in a haystack.

There is no cure for hemorrhagic colitis, and in fact, some forms of antibiotic therapy have been shown to increase the likelihood of kidney failure. Antidiarrheal agents should also be avoided because they will only serve to prolong the length of time that bacteria remain in the body. Supportive therapy and the body's own defense systems must battle invasion of *E. coli* O157:H7, and most healthy adults will recover on their own in five to ten days, while those who develop complications very often require intensive care and supportive therapy.

Shigella

Clinically, dysentery associated with consumption of fecally-contaminated water and sometimes food will yield a preliminary diagnosis that can be confirmed by finding *Shigella* organisms in the patient's stool samples. Biochemical tests will differentiate it from *E. coli* and other pathogens.

Those with mild infections will recover without antibiotic treatment, but more serious cases can be treated with certain antibiotics. Such treatment will shorten the duration of the illness. As with *E. coli* O157:H7, antidiarrheal agents should not be used. Other care is usually of a supportive nature, such as replacement of fluids and electrolytes. Even after a patient has recovered, bowel habits may not be normal for several months.

If there is a positive aspect of a *Shigella* infection, it is that for a period of up to several years, recovered individuals will be immune to reinfection, but only by the same species of *Shigella*.

Preventative Measures

E. coli

The most obvious preventative measures include the thorough cooking of ground beef to an internal temperature of 160°F or more, scrupulous washing of fresh fruits and vegetables, and the prevention of cross-contamination. To that end, consumers and food service personnel must be educated in safe food handling practices to help keep hazards to a minimum. Hand washing will prevent person-to-person spread, and is especially important among toddlers in daycare facilities, where *E. coli* infections can be difficult to control. Young children may continue to shed bacteria for up to two weeks after diarrhea has subsided, while older children tend not to harbor the bacterium in the absence of symptoms.

According to the CDC, other measures are being investigated for use on cattle farms and in beef processing plants, including the modernization of inspection practices and more sanitary methods of farming and slaughtering. Irradiation of beef would virtually eliminate most pathogens, including O157:H7. Methods are also being developed to decontaminate alfalfa spouts and seeds, and to prevent contamination of other types of produce. Other vehicles and strains of enterohemorrhagic *E. coli* must also be identified, and an international network for communicating information needs to be developed.

Researchers in the U.S. have developed a strategy to use bacteriophage (viruses that infect bacteria) to battle O157:H7 infection in cattle. Phage that are pathogenic for this bacterium could be incorporated into cattle feed, and once ingested, would hopefully destroy these deadly germs prior to slaughter, and before they have a chance to contaminate beef supplies. By studying the ecology of *E. coli* O157:H7, the CDC hopes to decrease its presence in food animals, learn how it contaminates produce, and eliminate its presence in both.

Computerized feeding of beef animals has supposedly played a role in the emergence of this pathogen as has the use of manure fertilizers, and inclusion of antibiotics in animal feeds, which has been done since the 1970s. In a unique twist of fate, it has been discovered that cows immune to brucellosis may also be immune to infection by O157:H7. In most parts of the country, this is a positive thing, but in the Northwest region of the U.S., the incidence of brucellosis has been greatly reduced and cattle there seem to be more prone to O157:H7 infection.

Probably the biggest problem contributing to the spread of this deadly pathogen is the way hamburger is processed, with the co-mingling of thousands of pounds of meat from countless animals—a single infected cow could contaminate 16,000 tons of ground beef. This practice makes it almost impossible to trace sources of *E. coli* O157:H7, and finding it in a single sample of ground beef necessitates the recall of thousands of pounds of product. As for those that prefer their hamburgers on the rare side, if you are a healthy (not elderly) adult, go ahead and live dangerously if you wish, but be sure that the beef your children, parents and grandparents consume is well done.

Shigella

In developed countries, the best way to avoid shigellosis is hand washing, although if a food has been contaminated by a handler, there is really no way to avoid ingesting these bacteria. Therefore, infected people or carriers should not prepare food, or even pour water for others. Travelers should take the same precautions already discussed with regard to other water-borne pathogens, such as *V. cholerae* and *E. coli*. In underdeveloped countries, proper sewage disposal and water treatment would drastically reduce bacillary dysentery and other diseases. Currently there is no immunization for shigellosis, but the government is involved in the development of such a vaccine. The Food and Drug Administration has developed a genetic probe to detect the virulence plasmid of *Shigella* bacteria in foods. The Environmental Protection Agency regulates and monitors the safety of drinking water in the U.S.

Potential Use as Agents of Food or Water Sabotage

Shigella does have a brief history of bioterrorist use. During the mid-1960s, there were several outbreaks of typhoid fever and dysentery in Japanese hospitals. The illnesses were traced back to food that had been intentionally contaminated by a research biologist, who later infected his family and neighbors. The Rajneeshee cult, responsible for contaminating salad bars in Oregon with *Salmonella* in 1984, had ordered but never used a culture of *Shigella dysentariae*.

In an isolated incident in Texas, *Shigella dysentariae* was the agent of intentional food sabotage in October 1996. Laboratory workers in a Texas medical center were anonymously "treated" to donuts and muffins that had been intentionally laced with this bacterium, which had apparently been stolen from the lab's stock freezer. Twelve of forty-five lab technicians who ate the pastries became ill with severe diarrhea, eight had *S. dysentariae* in their stool samples, eight required intravenous fluids, and four needed to be hospitalized. There were no deaths, and victims recovered with the help of antibiotics. *S. dysentariae* was also isolated from one of the remaining muffins. Neither the motive, the method of contamination, nor the saboteur were ever discerned, but investigators believe it was someone who knew what they were doing, including how to propagate the bacteria and effectively inoculate the bakery goods.

Unlike the bacterial pathogens discussed thus far, *E. coli* O157:H7 and *Shigella* species do not have long histories of biological weapons research, but both do have the potential to cause serious illness and sometimes even death when ingested. *E. coli* O157:H7 has not has not yet been known to have been used to sabotage food or water, but accidental contaminations of municipal water supplies have indicated that this organism can, under the right conditions, survive the levels of chlorine typically used to treat drinking water. Because it is a common soil and intestinal tract organism, intentional food or water contamination with pathogenic *E. coli* would appear to be accidental.

Neither *Shigella* nor *E. coli* O147:H7 is capable of causing death in healthy, non-elderly adults. Both can be destroyed by heat and disinfectants, and shigellosis can be treated with antibiotics. Once again, familiarization with the symptoms of infection and early medical intervention are important initiatives that individuals can take to protect themselves against these and other pathogens.

Escherichia coli and Shigella

Disease	Symptoms	Transmission	Prevention / Treatment	Food / Water Sabotage
Hemorrhagic colitis (*E. coli* O157:H7)	Severe cramps, watery or bloody diarrhea, sometimes vomiting	Usually undercooked hamburger; also unpasteurized apple juice & other foods	Avoid undercooked hamburger & unpasteurized apple juice; safe food handling; wash fruits & vegetables / Supportive	Possible / Possible (has happened naturally but need high numbers)
Gastroenteritis & traveler's diarrhea (Other types of pathogenic *E. coli*)	Various combinations of diarrhea, cramps, vomiting, fever	Food or water contaminated with fecal material (post-processing or foods that are not heated)	Avoid cross-contamination; cook foods thoroughly; avoid contaminated drinking water / Replace fluids & electrolytes	Possible / Possible (but need high numbers)
Shigellosis (*Shigella* species)	Bloody diarrhea	Fecally-contaminated water; fecal-oral route; foods contaminated by infected handler	Hand washing; infected people should not handle food; avoid contaminated water / Replace fluids & electrolytes; antibiotics	Possible (has been done before) / Possible but not likely due to dilution & chlorination

PART TWO

Nonbacterial
Agents

CHAPTER NINE

Food-borne Viruses and Other Viral Agents

This chapter deals with an entirely different kind of pathogen—viruses. Viruses can be even more frightening than bacteria because they cannot be battled with antibiotics; the human immune system is the only force able to fight viral infection. There are a few vaccines available for viral illnesses, but some, such as those used to guard against smallpox or polio, are no longer routinely administered.

Most viruses on the list of potential terrorist weapons are normally transmitted person-to-person either by sneezing or coughing, or by the bite of a mosquito. Experts on bioterrorism agree that these viral agents would be probably be dispersed as aerosols, though some have been known to be naturally transmitted via contaminated food or raw milk.

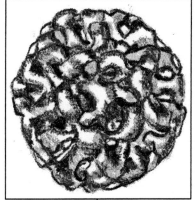

Norwalk virus

Characteristics of Viruses

Viruses are strange entities, existing on the very fringes of what scientists consider "alive." They are obligate intracellular parasites (must be inside a host to replicate), consisting only of genetic material (RNA or DNA) surrounded by a protein coat, and some have an outer layer of lipid (fat). Viruses are very specific with regard to the types of cells they infect, and when they are not inside a host, whether animal, plant or bacterial, they are inert and cannot reproduce.

Virus particles are extremely small and can be seen only with an electron microscope. The largest virus is fifty to one hundred times *smaller* than the smallest bacteria; about fifty million cold virus particles could literally fit on the head of a pin. *Germs* a book recently written by Judith Miller, Stephen Engelberg and William Broad, provides this analogy: if bacteria were the size of cars and minivans, viruses would be the size of cell phones.

Viral replication takes place only inside the cells of a host. All viruses reproduce by taking over the host cell's enzymatic machinery and using it to replicate themselves, destroying the host cell in the process. Hundreds of millions of new virus particles are released when the host cell bursts. These particles go on to infect additional cells in the host organism. New viral progeny inside the original host cell can also escape by squeezing through the membrane, but they take a bit of the membrane with them so this process eventually destroys the host cell as well.

Symptoms of viral infections run the gamut of severity, from brief periods of discomfort to severe manifestations that may end in death. Conversely, infection may not result in illness at all. For example, mononucleosis, caused by the Epstein-Barr virus, is carried by nine of every ten adults who never develop symptoms. A few viruses are on Centers for Disease Control and Prevention's list of emerging infectious diseases. These include the agents of AIDS, Rift Valley and Lassa fevers, and Ebola and Hanta viruses.

Smallpox and Polio Historical Perspectives

The existence of viruses as disease-causing entities was proven through the work of a Russian microbiologist, Dimitri Ivanovsky, who in 1892, discovered that something even smaller than bacteria was responsible for an illness in tobacco plants. The term "virus," which is Latin for poison, was used in the late 1890s to describe these elusive disease agents, but viruses themselves were not discovered until the early 1900s. Around 1930, W. M. Stanley, an American biochemist, purified Ivanovsky's tobacco mosaic virus.

Only two viral illnesses, polio and smallpox, have had significant implications in history. Egyptian carvings and paintings of four thousand years ago depict what seem to be victims of polio; the history of smallpox is over three thousand years old.

The introduction of smallpox to Hispaniola (now Haiti and the Dominican Republic) by Spanish explorers resulted in the deaths of over 3.5 million indigenous people, necessitating that slaves be brought from Africa to work on Spanish plantations. Smallpox also aided Cortes in his conquest of the Aztec empire. By the 1400s, the Chinese had already developed a method of immunization against smallpox, even though the causative agent was not yet known. Their early attempts at smallpox inoculations, however, resulted in full-blown disease in about one of two hundred persons.

These primitive vaccines were made from smallpox lesion exudates until 1796, when British physician Edward Jenner realized that milkmaids, who were constantly in contact with cows having the bovine version of the disease (cowpox), seemed to be immune from smallpox. This was easily observed because milkmaids were free of the scarring and blindness so often suffered by others who had survived the illness. Jenner hypothesized that the constant exposure to cowpox somehow protected these women from smallpox. To prove his theory, he inoculated an eight-year-old boy with cowpox then exposed him to smallpox. The boy remained healthy. Thus Jenner demonstrated that a disease that did not infect humans could indeed be used to vaccinate against a similar one that did. In fact, "vaccine" comes from *vacca*, the Latin word for cow.

Smallpox has continued to be an important cause of illness and death in developing countries until quite recently. The final epidemic occurred in Somalia in 1977, and the last reported cases were among laboratory workers in 1978. By 1980, the World Health Organization announced that smallpox had been eradicated through widespread immunization programs.

The first polio vaccine, developed by Jonas Salk of the University of Pittsburgh, was declared safe and effective in 1955. Albert Sabin of the University of Cincinnati developed the oral polio vaccine, which was approved for use in the United States in 1961. Although polio and smallpox have been eliminated as naturally occurring diseases, other, newly discovered viruses now present problems.

Smallpox as a Biological Weapon

The potential use of smallpox in biowarfare is particularly frightening. Since the disease was declared eradicated over twenty years ago, routine vaccinations are no longer administered as of 1972, though that may change in the very near future. Smallpox has also been used as a weapon in the past—in the mid-1700s during the French and Indian War, British soldiers gave blankets tainted with the virus to unfriendly Indian tribes. The Soviet Union reportedly maintains stocks of the virus today.

The World Health Organization is concerned because the smallpox virus is easily freeze-dried and capable of retaining its virulence for months or years in this form. These powders could initially be dispersed as aerosols, and by its nature, the disease would then spread person-to-person by coughing. Thus, lack of routine vaccination, the ease of large-scale production, and a high rate of infectivity make this virus one of much concern. An outbreak of smallpox in Europe in the 1970s showed how easily smallpox is transmitted through a population; a single individual infected eleven people, each of whom infected thirteen more.

When inhaled, the smallpox virus replicates within the lymph nodes of the respiratory system. Two to three days following the initial infection, a rash forms on the face, arms and hands, and lesions form in the mouth and pharynx. The illness then progresses downward, with the formation of lesions on the trunk and lower extremities. These lesions eventually become pus-filled and will scab over within two weeks. The scabs will fall off to leave indented, non-pigmented scars. The virus travels to other lymph tissue, the spleen, bone marrow, the lungs and the liver, producing malaise, fever, vomiting, backache and headache. The virus will also multiply in the bone marrow and spleen, which leads to a toxemic condition, and possibly encephalitis and death. Fifteen percent of victims will become delirious, and one third will die.

A study done at Yale and the Massachusetts Institute of Technology predicts that if an outbreak of smallpox in a large city affected one thousand people, within one year, those originally infected would pass the illness on, and eventually 110,000 people would die. If, however, just forty percent of the U.S. population were to receive the vaccine prior to an attack, only 440 would die. In September 2002, the CDC issued guidelines to the states indicating procedures for responding to an outbreak of smallpox, particularly how to vaccinate thousands of people in a very short amount of time—one million people in ten days. The CDC states that a single case of smallpox would be indicative of an intentional release and could trigger the widespread vaccination program if the government deemed it necessary. Vaccination prior to exposure or within two or three days after exposure to the smallpox virus will offer almost complete protection, while administration of the vaccine four or five days after exposure may protect against death, but not the development of symptoms.

Original plans earlier in 2002 were to vaccinate select individuals, those at high risk, in order to assess the efficacy of reintroducing the vaccine. But growing concerns prompted federal officials to take more action at an earlier time. Therefore, in September, the Bush administration began considering reintroduction of the smallpox vaccine to the general public in a pre-attack immunization program. This may be considered a drastic action since the government has not advocated use of the vaccine for the general public in the past, because it consists of the viable virus itself and carries some significant health risks—one death per million, and side effects in twelve people per million, including encephalitis and brain swelling. Therefore, whether or not to vaccinate the public is an area of current controversy.

Another disturbing bit of information came to light in July of 2002, when scientists at the State University of New York at Stony Brook constructed the polio virus from scratch using blueprints from easily obtainable, published data regarding the virus' structure. The project was undertaken simply to prove that such information can be used, possibly by terrorists, to create a viral bioweapon. The fear among these scientists and others is that other, deadlier viruses such as smallpox, could also be created this way, though the smallpox virus is much larger and more complex than the polio virus.

Other Potential Viral Agents

Venezuelan Equine Encephalitis

Venezuelan equine encephalitis (VEE), which is maintained in horses and other equines and transmitted by mosquitoes, was first isolated in 1936 in Venezuela, and initially documented as being acquired by humans in 1952 in Colombia. VEE is usually an acute illness (one with a sudden onset and short duration), but sometimes, more commonly in children, it may develop into encephalitis (inflammation of the brain). VEE has also been contracted by lab personnel via aerosols, and since its original isolation, over 150 such lab infections have been reported.

The Venezuelan equine encephalitis virus is transmitted by mosquitoes. It is indigenous to Central and South America, Mexico, Trinidad and Florida, where it is maintained in equines. In naturally occurring epidemics, illness among horses and other equines precedes human disease. VEE causes encephalitis, an inflammation of the brain. Seizures and fever are the predominant symptoms, and severe cases may lead to coma and death, though in less than one percent of infected individuals. The illness has a sudden onset of severe headaches, malaise, fever, muscle pain, extreme light sensitivity, and gastrointestinal symptoms. Most people recover within two weeks, and infection confers long-term immunity. Though it is seldom fatal, permanent neurological damage may result from this infection.

West Nile Virus

The West Nile virus is most commonly found in Africa, West Asia and the Middle East, and is closely related to the St. Louis encephalitis virus, which is indigenous to the United States. West Nile has just recently made its way to the Western Hemisphere as well, appearing in the U.S. at least by the spring of 1999, possibly earlier. Though it is not known where the U.S. virus came from, it is most closely related to the strains of West Nile virus found in the Middle East. The CDC states that its continued trek across the nation, from the East Coast in 1999 to the Midwest by 2002, indicates that it is here to stay. By October of 2002, West Nile had infected two thousand and killed ninety-eight people in thirty-two states.

West Nile infects humans, other mammals, and birds, and is transmitted through the bite of an infected mosquito. Even if you are bitten by a West Nile-carrying mosquito and subsequently become infected, there is a less than a one percent chance that you will become severely ill. In one of five infected people, it produces only a mild flu-like illness—fever, head and body aches, sometimes a rash on the trunk of the body, and swollen lymph nodes—that lasts a few days and yields no long-term health problems.

Severe illness occurs in one of every 150 infected individuals, primarily among those over the age of fifty-five. The virus multiplies in the blood, crosses the blood-brain barrier, and interferes with the functioning of the central nervous system. This illness manifests itself as encephalitis, meningitis or meningioencepahilitis, an inflammation of the brain and its surrounding membrane. Symptoms include a high fever, severe headache, a stiff neck, convulsions, disorientation, muscle weakness and stupor. Infection may progress to cause paralysis, coma and death. Recovery from the severe form of West Nile infection is possible, though symptoms may persist for several weeks and some neurological damage may be permanent. The West Nile virus is not transmitted person-to-person or animal-to-person, though it is suspected that it may be transmitted with blood transfusions. Infection most likely results in life-long immunity, though this immunity may decline over time. It is also possible that immunity to other viruses, such as yellow fever, dengue fever or St. Louis encephalitis, may also protect against West Nile.

Anyone experiencing symptoms of severe illness should see a doctor immediately. Infection with West Nile virus is initially diagnosed based on patient symptoms, and confirmed with a blood test. Treatment is supportive, possibly requiring intravenous fluids, ventilation, and protection against secondary infections. Currently there are no antiviral drugs approved by the FDA for treatment of West Nile infections, but a vaccine may become available within three years. Public health experts indicate that as more people develop immunity and mosquito populations are controlled, incidence of West Nile infection will decline.

Hemorrhagic Fever Viruses
The members of another order of viruses cause viral hemorrhagic fevers. The first outbreak of viral hemorrhagic fever caused by the

Marburg virus was recorded in 1967 in Marburg, Germany. It affected over thirty people in Germany and Yugoslavia after they were exposed to African green monkeys that were carrying the virus. There have been eighteen reported human outbreaks since, resulting in about 1,500 cases. Most of these outbreaks have occurred in Africa as a result of exposure to tissues, blood or secretions of infected primates or humans, or by contaminated syringes. The natural reservoir of the Marburg virus is not known, but is believed to be an animal indigenous to Africa.

Another hemorrhagic fever virus, Ebola, made its debut in 1976. It was first recognized as the agent of an outbreak of viral hemorrhagic fever in Zaire (now the Democratic Republic of the Congo), and Sudan. It is named after the Ebola River Valley, located in Zaire, Africa. There was a second outbreak of Ebola in Sudan in 1979. Initially the Ebola virus is acquired by contact with an infected animal, but then it can spread to other people via blood, secretions, or dirty needles. A large outbreak in Zaire in 1995 resulted from a single index case and ultimately affected over three hundred people. There have also been confirmed cases in Gabon, the Ivory Coast, and Uganda. A single case in England was the result of an accidental needle stick in a laboratory worker. In another isolated incident, one person in Liberia had developed antibodies to the virus but did not become ill. To date there have been no cases of Ebola reported in the U.S.

Several other hemorrhagic fever viruses are found primarily in Africa. They include Lassa, Rift Valley, Hanta, and Congo-Crimean viruses. Congo-Crimean fever is a tick-borne viral infection that is also found in the Crimea, Europe and Asia. Hanta virus was found in Manchuria before World War II, and later in China, Korea and Japan. Milder versions of Hanta virus exist in the Americas and in Europe. In the Southwestern U.S., this virus is carried by deer mice, and causes a debilitating respiratory illness that is fatal for twenty-five percent of its victims. An outbreak in the Four Corners region of the U.S. in 1993 resulted in thirty two deaths.

In 1977, a large epizootic epidemic of Rift Valley fever occurred in Egypt as mosquitoes transmitted the virus from animals to humans. The same thing happened in the Arabian Peninsula in 2000. There are several species of mosquitoes in the U.S. that are capable of spreading Rift Valley fever, which has also been transmitted via aerosols, contaminated food and raw milk. Lassa fever has also been spread through consumption of foods or by inhaling aerosols contaminated with rodent urine or excrement.

Viral hemorrhagic fever is a disease manifested by nonspecific symptoms of fever, headache, nausea, abdominal pain, diarrhea, muscle and joint aches, a skin rash and inflammation of the brain (encephalitis). As the illnesses progresses, patients may show signs of hemorrhaging within the body, and central nervous system disorders such as convulsions or delirium. Viral hemorrhagic fevers have an incubation period between two and twenty-one days, and last less than one week. Recovery is possible, but may be complicated by the inability to eat, weakness, fatigue, and inflammations throughout the body.

The symptoms exhibited by an infected individual will vary somewhat, depending on the agent. Death rates are also greatly dependent on the particular virus—the Omsk virus only kills one in two hundred of its victims, but a particularly virulent serotype of Ebola leaves ninety percent of its victims dead. Death from Ebola occurs within one to two weeks, ultimately due to hemorrhaging, intravascular coagulation, circulatory shock, and multi-organ system failure. (See reference table on following page showing the four classes of viruses that cause VHFs).

Other Viruses Spread Via Contact

Colds and Flu

Forty percent of the cases of the common cold are caused by rhinoviruses. These viruses are spread very easily through respiratory droplets when infected individuals cough and sneeze. Influenza is also a viral disease that is spread in the same way, but it is more serious than colds, causing over 100,000 hospitalizations and about 20,000 deaths each year in the United States.

AIDS

The most feared of all viral diseases is AIDS, or Acquired Immune Deficiency Syndrome. This virus is most commonly spread through sexual contact or shared syringes, but other means of transmission have been documented, particularly among people in medical professions. It is not the AIDS virus, however, that causes death, but rather what it does to the

Classes of Viral Hemorrhagic Fever Viruses

CLASS	MEMBERS	SPREAD VIA	DEATH VIA
Filoviridae			
	Ebola, Marburg	Direct contact with blood, secretions, tissues of patients or primates; contaminated needles; mucosal exposure. Ebola–slight chance of airborne transmission.	Necrosis of visceral organs, (liver spleen, kidneys); impairment of microcirculation
Arenaviridae			
	Lassa, New World Arena viruses	Inhalation of aerosols from, or contact with, urine and feces of infected rodents (abraded skin or mucous membranes); ingestion of food contaminated with rodent excreta; direct contact with blood or body fluids of infected individual; possibly airborne secretions.	Hemorrhage due to inability to form blood clots
Bunyavirirdae			
	Rift Valley Fever, Hanta	Bite of infected mosquito; direct contact with infected tissue; aerosols from infected carcass; ingestion of contaminated raw animal milk; lab aerosols	Blood vessel inflammation and liver tissue death
Flaviviridae			
	Yellow Fever, Omsk, Kyasanur Forest Disease	Bite of infected mosquito (yellow), bite of infected tick (Omsk, Kyasanaur); aerosol inhalation in laboratory	Encephalitis; degeneration of large visceral organs, especially liver and spleen; hemorrhagic pneumonia

immune system of its host. The AIDS virus attacks certain cells of the immune system, rendering them useless against common viral and bacterial pathogens, so AIDS patients succumb to diseases that ordinarily do not kill, such as diarrhea, pneumonia or various types of infections. AIDS cannot be contracted via food or water.

Food-borne and Water-borne Viruses

Viruses that are normally found in food and water cause an illness known as *viral gastroenteritis*. Viral gastroenteritis is primarily a problem for children, especially those under the age of two, and is a killer of millions of children each year worldwide. The most common agent of childhood viral gastroenteritis is rotavirus, which causes 3.5 million cases of diarrhea among children in the U.S. per year, and approximately thirty-five percent of hospitalizations for diarrhea in children younger than five. Worldwide, there are an estimated 140 million cases of rotavirus infection annually, resulting in four to ten million deaths. Rotavirus infects practically all children by the time they are four years old.

Norwalk virus affects primarily older children and adults, causing about forty percent of the cases of nonbacterial diarrhea in these populations. Almost seventy percent of adults in the U.S. possess serum antibodies to this virus. It was first recognized as the agent of an outbreak at a school in Norwalk, Ohio, in 1969, hence its name. Norwalk and similar agents are found primarily in seafood. Between 1976 and 1980, the CDC reported that half of Norwalk outbreaks were due to contaminated shellfish. Almost eight hundred people were affected in thirty-three separate outbreaks in England in 1976 that were associated with cockles. In Australia in 1978, a continent-wide epidemic affecting 2,000 people was linked to consumption of oysters. Over one thousand people became ill in a large outbreak in 1982 in New York state due to consumption of steamed clams. Salads were the second most common vehicle in Norwalk outbreaks from 1976 to 1980. Other foods that can also serve as vehicles for viruses are typically those that are not heated prior to being consumed. If such foods become contaminated with Norwalk virus, infected food handlers are most commonly implicated. Between 1985 and 1988, nine of the fifteen documented Norwalk outbreaks reported to the CDC were due to infected food handlers.

The most common cause of viral gastroenteritis among adults is hepatitis A. Between 1938 and 1989, its incidence increased in the U.S. by eighty-five percent, going from almost ten to fifteen cases per 100,000 people annually. Although an outbreak of hepatitis A may start with contaminated food or water, it is perpetuated by person-to-person transmission. For example, an outbreak in Shanghai originally affected over 16,000 people who ate raw shellfish, but person-to-person spread of the virus ultimately caused illness in an estimated 300,000.

The common cold is the number one cause of human illness, and viral gastroenteritis takes the number two spot. The most common food-borne viruses are hepatitis A, Norwalk and Norwalk-like viruses. The most common vehicle for these viruses is shellfish harvested from water that is contaminated with human waste. In their aquatic habitat, shellfish obtain food by "filter-feeding," forcing water through an internal filter system and extracting food particles. This process actually serves to concentrate viruses present in the water, sometimes as much as nine hundred times. Viral gastroenteritis is also often attributed to contaminated water from pools, lakes, municipal supplies, and even ice. Enteric viruses are transmitted person-to-person via the fecal oral route. This type of transmission occurs most commonly within institutional settings. Norwalk and rotavirus have also been transmitted by aerosols.

Viruses can be deposited into any kind of food by a handler, someone who may be recovering from a viral illness, or someone who is a carrier. The good news is that unlike bacteria, viruses cannot multiply in the foods they contaminate, so temperature abuse of foods is not a contributing factor in food-borne viral gastroenteritis. Viruses are typically destroyed by the heat of cooking. There have, however, been outbreaks involving steamed clams. Because of their protein shell, viruses are more resistant to adverse environmental conditions than their bacterial counterparts, and have been found to survive eight days in hamburger at room temperature.

Disease Manifestations of Viral Gastroenteritis

Viral gastroenteritis is typically a mild, self-limiting illness, usually involving nausea, vomiting, non-bloody diarrhea, fever, headaches and abdominal pain, though infections by the different types of viruses may result in additional symptoms such as muscle pain, chills, and sometimes even a sore throat. Incubation periods vary from one to ten days, with symptoms lasting three to seven days. Though the exact number of ingested virus particles needed to cause illness is unknown, it is presumed to be very low, perhaps just ten.

Viral Infections

Rotavirus

Most children in the U.S. become infected with rotavirus between the ages of six months and two years, and many outbreaks are associated with daycare centers. Children who attend daycare are more frequently infected with rotaviruses, as well as Norwalk, than are children who do not attend daycares. Rotavirus is easily transmitted by dirty hands, though it can survive outside a host for several days under conditions of low temperature and low humidity, and can be picked up from surfaces such as toilets. Rotavirus infections are most common during the cooler months of the year.

The illness begins twenty-four to seventy-two hours after ingesting virus particles. The initial symptom is vomiting, which is followed later by diarrhea. Symptoms last three to eight days, and children and older adults may need fluid and electrolyte replacement. In the U.S., child mortality is low, but worldwide, it is estimated that over one million children die each year from diarrhea caused by rotavirus. Primary infection leads to immunity from severe recurrence of infection, though individuals may still develop mild diarrhea.

Norwalk

Norwalk and similar viruses, also referred to as SRSVs or small round structured viruses, belong to a family wherein agents are named after the area where they were originally implicated in an outbreak. Norwalk virus is named after Norwalk, Ohio; other viruses in this family include Snow Mountain, Saporo, Taunton, Montgomery County, and Hawaii. Norwalk affects primarily school-age children, who suffer mostly with vomiting. Older adults are also susceptible, but their main symptom is diarrhea. Adult deaths have occurred from Norwalk infection due to electrolyte imbalance. Symptoms begin twenty-four to seventy-two hours after ingesting the agent, last one or two days, and are most commonly accompanied by abdominal pain and nausea, though may also include headaches, chills and muscle pain. Immunity is

135

rather limited, with re-infection possible after about two years. Shellfish and salads are the primary food vehicles of transmission, but other documented routes include aerosols, fomites (contaminated objects), and person-to-person transfer.

The risk of viral gastroenteritis, particularly caused by Norwalk, increases in areas of crowding and decreased personal hygiene, such as nursing homes and residential institutions. The elderly, the young and those with compromised immune systems, including pregnant women, are also at increased risk. (While diarrhea due to any agent increases risks during pregnancy, there is no particular risk associated with viral gastroenteritis.) Cruise ships and camps also present increased risk environments because of close living quarters which promote person-to-person spread of the virus, and a regular arrival pattern of a new, uninfected population every one or two weeks. One outbreak of Norwalk virus on a particular cruise ship lasted for five successive trips.

Outbreaks have also occurred in schools, at banquets, in fast-food restaurants, and recreational areas such as campgrounds and swimming pools. Norwalk viruses are also agents of traveler's diarrhea.

Hepatitis A

The hepatitis A virus is the most common cause of adult viral gastroenteritis. There are several types of hepatitis viruses, but the majority of food-borne hepatitis outbreaks are due to type A (previously known as infectious hepatitis), and type E. All forms can also be spread via contaminated water, blood, sexual contact, or shared syringes.

The word "hepatitis" comes from the Greek word for liver, *hepato*, and the hepatitis A virus causes an inflammation of the liver. Symptoms of hepatitis can range from mild and flu-like to severe, and infection may even lead to liver failure depending on the victim's immunity. The hepatitis A virus occurs worldwide. Outbreaks can be sporadic or epidemic, and surprisingly are more common in developed countries. This is due to the fact that people in developing countries are exposed to the hepatitis virus from an early age, and thus acquire immunity to it very quickly.

Infection with this virus is most common among young people, and is mainly spread by the fecal-oral route. Those considered to be at risk for contracting hepatitis A include individuals who come into contact with hepatitis patients on a regular basis, such as those working in medical facilities or daycare centers, as well as frequent international travelers. Hepatitis A outbreaks have occurred in crowed areas such as nurseries and prisons. It can also be spread among drug users sharing needles, but AIDS is a much more serious consequence of this careless practice.

Foods can also transmit hepatitis A, and there are more documented outbreaks of hepatitis A associated with food than any other virus. One outbreak due to contaminated strawberries imported from Mexico affected primarily school children. Imported lettuce, iced tea, raw oysters and ice-slush drinks have also served as vehicles.

Incubation times for hepatitis infection vary from two to six weeks. Symptoms start abruptly with loss of appetite, nausea, fever and malaise, though early signs may also include sensitivity in the upper right abdomen, followed several days later by jaundice. There may also be generalized itching, swollen lymph nodes, and intermittent diarrhea. Gastrointestinal symptoms last only ten days, but full recovery can take as long as eight weeks. Hepatitis has a relapse rate between ten and fifteen percent, though relapse is usually due to poor patient judgment, such as trying to resume normal activity levels too soon, or to alcohol consumption, which further taxes an already-weakened liver. Infection yields lifelong immunity, and sixty-five percent of Americans over the age of fifty are immune to hepatitis A. Fatalities are very infrequent.

Controlling Viral Gastroenteritis

Regardless of the original source of the virus—undercooked or raw shellfish, other foods contaminated by a handler, or contaminated water—viral gastroenteritis is usually perpetuated via the fecal-oral route, and sometimes by aerosols. Because of this, improving water and food sanitation may have only a minimal effect in controlling outbreaks of viral gastroenteritis. The source of the virus must be eliminated, so people who are recovering from a diarrheal illness should not handle food. Norwalk viruses can remain infective even after freezing or heating to 140°F for thirty minutes. However, all viral agents of gastroenteritis, including Norwalk, should be destroyed by boiling for ten minutes or more.

Exposure of susceptible populations—the very young, the very old, pregnant women and the immunocompromised—should be avoided. To prevent person-to-person spread in daycare or health care environments, recovering individuals should not be in direct contact with susceptible persons for at least two days after their symptoms subside. Employees in retirement homes, daycares, hospitals, etc. should also take measures to protect themselves against viruses and other agents by wearing gloves and, if necessary, gowns and masks since some viruses can be spread in the air.

As always, one of the most important preventative measures is thorough hand washing with soap and water for a minimum of thirty seconds. Norwalk, rotavirus and hepatitis A, however, will actually survive ordinary hand washing. Bleach or chlorine is necessary for their inactivation. Most household cleaning agents will destroy these viruses.

Care should be taken in disposing of soiled clothes and linens, and in cleaning and sanitizing of surfaces. Though chlorine bleach will destroy viruses and bacteria, contact time is important. Surfaces should be washed with soap, then rinsed with diluted bleach solutions and allowed to air dry to optimize the amount of contact time that the sanitizing agent has to destroy the virus particles. Norwalk viruses are able to withstand thirty minutes of exposure to 6.25 milligrams of chlorine per liter of sanitizing solution. It actually requires ten milligrams of chlorine per liter for inactivation. Because of this resistance, Norwalk is implicated in about twenty-five percent of outbreaks associated with municipal drinking water.

The risk of death from viral gastroenteritis is highest among the malnourished living in areas where medial care is not readily available. Vaccination against prevalent viral agents, such as rotavirus, is the best way to control infection. Trials of rotavirus vaccines are currently underway.

CDC's Priority List for the Control of Intestinal Viruses

The Centers for Disease Control and Prevention has identified three major goals for viral research efforts: improvement of diagnostic capabilities for the known viral pathogens to aid in determining their endemic importance and their role in outbreaks; identification of new agents for the fifty percent of diarrhea outbreaks for which the agent remains a mystery; and determination of the mode of transmission and the means to prevent viral infection, including the characteristic of natural immunity and effective vaccination programs.

Diagnosing Viral Illness

Antigen kits are currently available for many viral agents, and are under development for others. Such kits have been used for doing research on Norwalk, Snow Mountain, and other viruses. Serum antibody levels in a patient's blood can be used to diagnose a viral infection, but antibodies do not appear until the period of recovery, usually at the second week. To confirm infection by viral hemorrhagic fever viruses, samples must be sent to the CDC or the U.S. Army Medical Research Institute of Infectious Diseases (see Resources). However, eventually, select U.S. public health labs will be able to perform this testing, which is done using PCR, ELISA, or viral isolation.

In the case of viral gastroenteritis, virus particles can be identified in patient stool samples with the help of antibodies and an electron microscope. These specific antibodies cause the viruses to agglutinate (stick together) so they are easier to locate in the microscopic field. Polymerase chain reaction (PCR) is currently available only for rotavirus, but is being developed for other viruses. Viruses are difficult or impossible to cultivate, and are therefore not grown routinely in diagnostic laboratories. Rotavirus, enteric adenoviruses, and astroviruses can be cultivated, but Norwalk and others cannot. Viruses are also difficult to detect in environmental and food samples, though some success is seen utilizing mammalian cell culture infectivity assays, or immunofluorescent staining. In order to perform these assays, however, it is first necessary to concentrate the virus particles, a task that is difficult at best.

Preventing and Treating Viral Infections

Vaccines are available for some hemorrhagic fever viruses—Bolivian, Argentine, Lassa fever and Crimean-Congo fever—though the only licensed virus-specific vaccine is that for yellow fever. Unfortunately, this vaccine is produced only in limited supply, and would not be useful if administered after an attack because yellow fever has a short incubation period (three to six days), and antibodies would not be formed before illness ensued. Other vaccines are undergoing development and testing, including those for Rift Valley fever, Hanta virus and dengue fever.

Antibiotics are ineffective against viruses, so humans must rely on their immune systems to battle viral infections. Some antiviral drugs may, however, help to control viruses. For example, acyclovir used to treat herpes simplex and AZT for treatment of AIDS. Both inhibit viral replication and thus hold these viruses in check. There are a few drugs available to treat viral influenza, but they must be given very early in the course of infection, within the first forty-eight hours. Otherwise, treatment is limited to controlling only the symptoms of viral illness—replacing fluids and electrolytes, or treating pain and fever. Severe cases of viral illness, such as those associated with hemorrhagic fever viruses, may require mechanical ventilation, dialysis, or anti-seizure medications. Currently there are no antiviral drugs approved by the FDA for treatment of infection with hemorrhagic fever viruses.

Viruses as Biological Weapons

Viruses are attractive as weapons because there are few ways to treat viral infections, they are easily transmitted, and they have the potential to cause illness and sometimes death. According to the Working Group on Civilian Biodefense, the former Soviet Union and Russia have produced significant quantities of Marburg, Lassa, smallpox and Ebola viruses. In the 1950s, the United States developed encephalitis, yellow fever, smallpox, and Rift Valley fever. There are also reports that yellow fever may have been weaponized by North Korea. The Venezuelan equine encephalitis virus was weaponized by the United States in the 1950s and 1960s. Compared to other viruses, VEE and similar viruses are easier to produce in the lab in significant quantities and using inexpensive and unsophisticated equipment, are stable in storage, and infective in aerosol form.

The most dangerous viruses, and therefore those most likely to be made into bioweapons, are those that can be effectively transmitted through the air. The Working Group on Civilian Biodefense has identified the following agents of viral hemorrhagic fevers to be of concern: Ebola, Marburg, Lassa fever, New World Arenaviruses, Rift Valley fever, yellow fever, Omsk hemorrhagic fever, and Kyasanur Forest disease. But in their natural state, these viruses do not persist for long in the environment.

Intentional Viral Contamination of Food or Water

Though it has been demonstrated that viruses can indeed be produced for use as biological weapons, those of choice would probably be released as aerosols. Polio was typically spread by contaminated water, and an epidemic around the time of World War I was due to raw milk. Polio is not one of the agents of current concern since vaccines are still routinely administered. According to a report from the Working Group on Civilian Biodefense, arenaviruses such as the Lassa fever virus have been known to be transmitted through foods contaminated with rodent excreta, and Rift Valley fever has been contracted by ingestion of contaminated raw milk. Thus it is possible to contract one of these hemorrhagic fever viruses via ingestion. It is unlikely, however, that these agents or agents of viral gastroenteritis would be used to contaminate food or water.

As mentioned, unlike bacteria, viruses are very difficult to grow in the laboratory, primarily because they only reproduce inside other living cells. These host cells have to be cultivated first, then infected with virus. Alternatively, fertilized chicken eggs may be used for viral propagation. In either case, cultivation of viruses is costly, time-consuming, and requires a high degree of technical expertise.

According to the Working Group on Civilian Biodefense, the viruses most likely to be used as biological warfare agents are those that cause hemorrhagic fever. The exact means by which these viruses are transmitted among the human population is still not fully understood, and most importantly, the true significance of airborne transmission must be determined.

There is also a need for readily available rapid diagnostic methods, a safe means to handle specimens in the lab, as well as vaccines and drug therapies. Vaccines available for investigational use include those for Argentine and Bolivian hemorrhagic fevers, Rift Valley fever, and Kyasanur Forest Disease. Vaccines under development include those for filoviruses (Marburg and Ebola) and Lassa fever. The antiviral drug ribovarin is FDA approved for the treatment of hepatitis C; this drug could eventually be approved for the treatment of viral hemorrhagic fevers as well. A vaccine offering the elderly protection against West Nile could become available within three years, and there is some evidence that immunity to other viruses such as yellow fever, dengue fever and St. Louis encephalitis may also protect against West Nile virus.

Smallpox is the virus of primary concern at this time. Vaccine is available, and as of this writing, the Bush administration is evaluating the ramifications of making it available to the general public.

Viruses

Disease	Symptoms	Transmission	Prevention / Treatment	Food / Water Sabotage
AIDS	High susceptibility to infections due to decreased immunity	Sexual contact or shared needles	Safe sexual practices; do not share needles / Anti-viral drugs	No / No
Hemorrhagic fever (Hanta, Ebola and others)	Fever, headache, nausea, abdominal pain, diarrhea, myalgia, joint pain, rash, encephalitis	Direct contact with infected tissues etc; bite of infected mosquito or tick; inhale aerosols; ingestion of contaminated raw milk (Rift Valley) or food (Lassa)	Avoid contact with infected animals etc. / Some vaccines available; supportive care	Possible but not likely / Possible but not likely (Lassa or Rift Valley fevers) (aerosol dispersal possible for all)
Viral encephalitis (VEE, West Nile)	Seizures, fever, headache; malaise, myalgia, gastrointestinal symptoms (VEE)	Bite of infected mosquito	Control mosquito populations / Supportive	No / No
Viral gastroenteritis	Nausea, vomiting, diarrhea, fever, headache, abdominal pain	Dirty hands or surfaces; raw shellfish; water contaminated with human fecal material; infected food handlers	Avoid contaminated water & raw or undercooked shellfish; wash hands; infected people should not handle food / No treatment	Not likely / Not likely
Smallpox (*variola*)	Rash and lesions starting on face and progressing downward	Person-to-person	Avoid infected persons / Vaccination if available	No / No (aerosol dispersal possible)

CHAPTER TEN

Natural Toxins

Some of the most deadly substances known are from Mother Nature herself. These include many bacterial toxins previously discussed, and the present chapter deals with toxins produced by plants, algae and

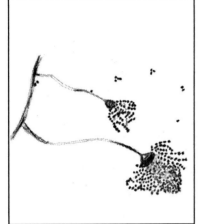

molds. These include ricin, a by-product of the manufacture of castor oil, marine toxins, and mycotoxins produced by various types of molds.

Some of theses natural toxins have been used or have been considered for use as biological weapons. Many of them are widespread in nature, and can be easily obtained and weaponized. They all have deleterious affects, and some can kill with relative ease.

Aspergillus

Ricin

Ricinus communis, the common castor bean, is found throughout the world and one million tons are used annually in the production of castor oil. Ricin is a stable, toxic by-product of this process. It comprises about five percent by weight of the waste material generated in the extraction of the oil from castor beans. Ricin has a brief but interesting history of use as a biological weapon. In 1978, in London, a Bulgarian dissident, Georgi Markov, was assassinated with a ricin-infused projectile disguised as an umbrella tip. Ricin was injected intramuscularly into Markov, who died the following day.

Ricin Intoxication

Naturally occurring ricin intoxication is due to ingestion of castor beans themselves, which produces severe gastrointestinal symptoms followed by collapse of the victim's blood vessels, and death. Ricin can also be fatal by inhalation or injection. Ricin is toxic to mammalian cells because it inhibits protein synthesis. If ricin toxin is inhaled, there is an abrupt onset of illness from eight to twenty-four hours after exposure, and includes respiratory symptoms of chest tightness and cough, fever, nausea, joint pain, and shortness of breath. Death due to blood vessel collapse can occur in two to three days, depending upon the dose. In the rare event of ricin poisoning by injection, muscle tissue at the injection site would die, and the victim would suffer intravascular coagulation, microcirculatory failure and possibly multiple organ failure. If ricin were to be used as a food or water contaminant, symptoms of ingested toxin would include severe gastrointestinal symptoms with intestinal bleeding and necrosis of spleen, kidney, and liver tissue. Once again, death would be attributed to vascular collapse.

Diagnosis and Treatment

If ricin were to be delivered as an aerosol, it would produce symptoms similar to many other types of inhaled biological agents, such as anthrax, *Staphylococcus* enterotoxin B (SEB), Q fever, or others. Ricin poisoning, however, would not respond to antibiotics, and would differ from SEB in that intoxication with SEB would plateau clinically and most likely not end in death of the victim. If ricin were ingested, it could be differentiated from enteric pathogens because ricin poisoning produces the unique symptom of vascular collapse. A suspected case of ricin poisoning could be confirmed by finding antibodies in the patient's blood, or by detecting castor bean DNA. Immunohistochemical techniques, as well as PCR and ELISA assays, are available for these purposes.

Treatment for inhaled or injected ricin toxin would be mainly supportive. However, if ingested, the gastrointestinal tract could be decontaminated with activated charcoal or other methods, and supportive therapy would include fluid replacement.

Preventative Measures

Ricin is not active on the surface of the skin. It is not volatile and there is no secondary aerosolization. Skin and clothing can be decontaminated with soap and water. In case of aerosol exposure, a mask would provide protection. Ricin is inactivated by chlorine, so a bleach solution can be used to decontaminate surfaces and objects that may have come in contact with this toxin.

Antitoxin is not available for ricin poisoning, nor is there a vaccine, though immunization of animals has been done successfully.

Ricin as a Biological Weapon

The worldwide distribution of castor beans and the fact that ricin comprises five percent of the waste material from the production of castor oil means that this toxin is readily accessible and easily obtained. It is also easy and economical to produce in several different forms—powder, liquid, crystalline—using simple equipment. However, it is much less potent than other toxins such as botulism toxin, so large quantities would be necessary to effectively poison a large number of people. This fact alone may deter a terrorist from using ricin as a weapon.

As with so many of the other biological weapons, terrorists would probably distribute ricin as an aerosol but because it is not a strong toxin, dissemination by air or water would not be particularly effective due to the diluting effects. The chlorine used in municipal water supplies would be enough to render the toxin inactive. Also, reverse osmosis, sometimes used in processing drinking water, would remove ricin. Of all the available methods of dispersal, contamination of food seems the most likely for this toxin, the only potential deterrent being the need for large quantities. Ricin could be used to contaminate any type of food, but since large amounts would have to be added, a saboteur would probably prefer to use something more potent, though ricin may be easier to obtain.

Marine Toxins

As the name indicates, marine toxins are found in the ocean. They occur naturally, are produced by certain types of bacteria and algae, and contaminate various kinds of seafood. Unfortunately, they cannot be detected by sight or smell, and are not destroyed by heat.

Types of Seafood Poisoning

Scrombroid: The most common seafood poisoning in the United States is *scrombrotoxic fish poisoning*, also known as histamine poisoning. Finfish such as tuna may undergo bacterial spoilage between the time they are harvested from the ocean and the time they reach your plate. These bacteria break down protein in the fish and produce histamine, which can cause severe allergic reactions when the fish are eaten. Within two minutes to two hours, victims will break out in a rash and suffer headache, sweating, diarrhea and vomiting, and may also experience abdominal pain and a burning, swelling, or metallic taste in the mouth. Most people will recover on their own within a few hours, but if symptoms are severe, epinephrine or antihistamines can be administered.

Ciguatera: The second most common marine poisoning, *ciguatera*, is caused by ciguatoxins produced by certain types of dinoflagellates that are eaten by tropical reef fish. These small fish are eaten by bigger fish, which are eaten by yet bigger fish, so the toxin becomes more concentrated as it moves up the food chain, reaching its highest levels in large predatory fish like barracuda, sea bass, or grouper. Ciguatera poisoning occurs most frequently from eating fish that have been caught on reefs in Hawaii, the South Pacific, Virgin Islands, and Puerto Rico.

Ciguatoxins cause nausea and vomiting, cramps and diarrhea, sweating, headaches, a pins-and-needles burning sensation, weakness, itching, dizziness and muscle pain within six hours of ingesting contaminated fish. Sometimes the victim will also experience the strange sensation of temperature reversal wherein hot foods seem cold and vice versa. There may also be other unusual tastes, or even nightmares and hallucinations. Ciguatera poisoning is not typically fatal, and symptoms will disappear in one to four weeks, though sometimes neurological problems may persist longer.

Paralytic Shellfish Poisoning: *Paralytic shellfish poisoning* is the third most common marine illness, and is also due to a toxin produced by a reddish brown dinoflagellate. When these organisms reach high levels in the summer, they form red bands of color in the ocean, causing what has become known as "red tides." Shellfish such as mussels, clams, oysters, lobsters, crabs, cockles and scallops living in cold coastal waters of the Pacific states and New England will concentrate this toxin in the process of feeding.

The incubation time of this poisoning can vary from fifteen minutes to ten hours, but symptoms usually start within two hours after eating contaminated shellfish. Victims will experience a mild facial numbness or tingling, followed by similar sensations in the arms and legs, headache and dizziness, nausea, drowsiness, incoherent speech, and muscle uncoordination. While most cases of paralytic shellfish poisoning are quite mild, if the dose of toxin is high enough, paralytic shellfish poisoning can be life threatening due to paralysis of respiratory muscles, with death occurring within two to twenty-five hours.

Neurotoxic Shellfish Poisoning: Yet another type of dinoflagellate produces *neurotoxic shellfish poisoning* when it becomes concentrated in shellfish from the Gulf of Mexico and the southern Atlantic states. Oysters, mussels and clams have been implicated in this type of poisoning, which has an incubation period of one to three hours and symptoms similar to paralytic shellfish poisoning, as well as tingling and numbness of the lips, throat and tongue, muscle aches, and sometimes the temperature reversal symptom manifested with ciguatera poisoning. Victims of neurotoxic shellfish poisoning typically recover in two to three days, and no deaths due to this type of poisoning have been reported.

Amnesic Shellfish Poisoning: The rarest form of marine poisoning is *amnesic shellfish poisoning*. This toxin is produced by the diatom *Nitzchia pungens*, a microscopic water plant. Shellfish concentrate the toxin as they feed on these diatoms. Ingesting the contaminated shellfish causes gastrointestinal symptoms within twenty-four hours, as well as neurological symptoms including headache, confusion, dizziness and disorientation. The name of this type of poisoning, however, comes from the bizarre symptom of permanent short-term memory loss. Death may occur in severe cases, and is preceded by seizures, focal weakness or paralysis, and coma. All recorded deaths due to amnesic shellfish poisoning have so far been among the elderly.

Diagnosis and Treatment for Marine Toxin Poisoning

If a patient shows characteristic symptoms of marine poisoning previously described, and has recently eaten a suspect type of seafood, diagnosis is usually based on just these observations. Any leftover food can be assayed for the presence of toxin.

For the most part, marine poisoning can only be treated with supportive care, though antihistamines and epinephrine may be used to ease symptoms of scrombrotoxic fish poisoning. Victims of paralytic shellfish poisoning may require respiratory support, which if given within twelve hours of eating contaminated shellfish, usually leads to complete recovery.

Frequency of Marine Toxin Poisoning

There are about thirty cases of marine poisoning reported annually in the U.S., the incidence increasing in warmer months along with the population of dinoflagellates. The Food and Drug Administration and the Centers for Disease Control and Prevention agree that the number of reported cases is most likely much lower than the actual number since health care workers are not required to report marine toxin poisonings, and mild episodes do not warrant medical attention. The CDC estimates that one person will die from toxic seafood every four years.

Preventative Measures

Fresh fish should be kept refrigerated and eaten as soon as possible to prevent development of histamines. The CDC recommends that barracuda not be consumed, especially if it is from the Caribbean. The CDC also suggests checking with health authorities before harvesting shellfish, especially during times of algal blooms or red tides.

Shellfish are tested in some areas by local health departments to monitor levels of toxins and devise control measures. State and federal agencies keep track of cases of marine poisoning and try to determine if a particular fishing area or shellfish bed has become contaminated.

Marine Toxins as Biological Weapons

Because of its potency and potential to cause death, the most likely of the marine toxins to be used as a biological weapon would be that which causes paralytic shellfish poisoning. Five grams of this toxin were found among the other agents retained by the Central Intelligence Agency after Nixon's ban on offensive biological weapons.

This toxin could be delivered as an aerosol or as a contaminant of food, but chlorine will inactivate these toxins on surfaces (ten percent bleach solution) as well as in municipal water supplies. As with ricin, aerosol or water delivery would be inefficient due to diluting affects.

There are no vaccines or antitoxins yet available, but poisoning due to ingestion of marine toxins could be treated either by inducing vomiting, or giving the patient activated charcoal, which absorbs the toxin from the gastrointestinal tract.

Mycotoxins: Toxins Produced by Molds

Many different types of molds have been shown to produce toxins that can affect plants, animals, humans or even microorganisms. These mycotoxins may cause mutations in DNA, may be carcinogenic (cancer-causing), or even toxic to specific organs. Aflatoxins, for instance, are toxic to liver tissue.

Aflatoxin: Aflatoxin poisoning was initially discovered in 1960 in England. Over 100,000 turkeys died as a result of what was called "Turkey X" disease. It was later discovered that the feed was contaminated with aflatoxin from Brazilian peanut meal.

Aflatoxins are produced by *Aspergillus flavus*, *A. parasiticus*, and some species of *Penicillium*. Most commonly, these molds grow and produce toxin in grain crops such as barley, corn, rice, soybeans and oats, and in peanut products. Aflatoxins come in several varieties with different toxicities. Some of these are produced by the molds themselves, while other varieties of aflatoxin are produced by animals when they eat grains that are contaminated with these mold-produced varieties. The latter types are secreted in the animal's milk.

Humans who ingest aflatoxins from milk or grain may suffer either short-term or long-term effects. The short-term illness, called *aflatoxicosis*, progresses rapidly and presents with initial symptoms of vomiting and anorexia, followed by fluid accumulation in the lower extremities and jaundice. In experimental animals, these toxins cause lesions in the liver, and large doses result in liver tissue hemorrhage and necrosis. An outbreak of aflatoxicosis in India in the 1970s had a mortality rate of almost thirty percent. Death came suddenly following massive gastrointestinal hemorrhage. The long-term effect of aflatoxin consumption is cancer of the liver.

Aflatoxin production can be curtailed in crops by controlling mold growth. This entails the use of good farming practices, prevention of crop damage, and possibly the use of chemicals. Molds will also grow during crop storage, so the environment must be controlled in terms of humidity, temperature and insect pests. Food products that are prone to aflatoxin contamination, such as peanut butter, must be stored properly under conditions of low temperature and humidity. The FDA regulates the aflatoxin content of foods produced in the U.S.

Ergot: Ergotism, or ergot poisoning, is caused by different kinds of mycotoxins. The first such toxin to be isolated, ergotamine, is a derivative of lysergic acid, the synthetic form of which is the hallucinogenic drug, lysergic acid diethylamide, or L.S.D.

Caused by *Claviceps purpurea* and other closely related species of mold, ergotism was a recurring problem in Europe for over two thousand years, evidenced by countless epidemics during the Middle Ages which peaked from 1770 to 1780 in France and Germany. Patients often developed gangrene accompanied by burning pains and the sensation of intense heat, so the illness was also known as "St. Anthony's Fire." In attempts to thwart the disease, pilgrimages were made to the relics of Saint Anthony. Many people lost fingers, toes or entire limbs due to gangrenous ergotism, and 40,000 died. Epidemics continued until 1855. The last outbreaks occurred in Russia and England in 1928.

Gangrenous ergotism was the most common form during the Middle Ages, but another form, convulsive ergotism, which predominated during the last few epidemics, made people appear to be bewitched as they suffered convulsions and hallucinations that alternated with periods of drowsiness. Death or permanent mental impairment was due to lesion formation on the brain and spinal column. Some individuals persecuted as witches in Massachusetts in 1692 may have been suffering from this type of ergotism.

Though the incidence of ergotism has greatly declined, it is still necessary to monitor susceptible crops such as rye or wheat to control mold growth and subsequent toxin production. Measures include the use of resistant crop varieties and certified seed, crop rotation, and the removal of mold from crops.

Other Mycotoxins

Citrinin: Citrinin, produced by several species of *Penicillium* and a few species of *Aspergillus*, is a toxin that breaks down and kills kidney tissue. It is found in various cereal grains and has been associated with yellow rice disease in the Far East.

Ochratoxin A: Ochratoxin is also produced by *Aspergillus* and *Penicillium* species and is found in wheat, barley, oats, corn, beans and peanuts. In animals, ochratoxin affects the kidneys and is believed to have the same effect in humans.

Patulin: *Penicillium expansin, P. clavicin* and several other molds produce this toxin, which causes mutations, malformations, and cancer. It is found in fruits and fruit products, particularly apple juice.

Zearalenone: This toxin, also called F-2 toxin, is produced by *Fusarium* molds and is associated with corn. It causes infertility, though this effect has only been seen in animals.

Trichothecene Mycotoxins (T2)

Trichothecene mycotxins are very small compounds produced by some types of filamentous (thread-like) molds. Similar to other agents discussed, they act by inhibiting protein synthesis at the cellular level, impairing DNA synthesis, and interfering with the structure and function of cell membranes. The main effects of these fast-acting toxins are on body tissues that are rapidly proliferating, such as bone marrow and skin.

Naturally-occurring tricothecene mycotoxins caused *alimentary toxic aleukia*, also known as septic angina, in certain areas of Russia prior to World War II. These toxins were contaminants of cereal grains—millet, wheat, rye, oats and buckwheat—that were customarily left in the fields over the winter and harvested the following spring. Climactic conditions such as mild winters, heavy snowfall, and a slow spring thaw contributed to production of these toxins.

Historically, alimentary toxic aleukia progressed in stages, the first of which was an inflammation of the mouth, throat and gastrointestinal tract. One to three days later, victims experienced nausea, vomiting and diarrhea that lasted a week or more.

After these symptoms disappeared, stage two ensued without exhibiting any outward signs, but during the next two-week to two-month period, bone marrow was being destroyed. At the end of this stage, small hemorrhages appeared on the skin. The final stage, which lasted five to twenty days, culminated in total atrophy of bone marrow, enlargement of skin hemorrhages, and the formation of necrotic lesions of the skin and muscles. Necrotic lesions in the mouth and throat became large, and in over thirty percent of cases, death was caused by swelling around the vocal cords (acute stenosis of the glottis). Sometimes the pulmonary system was also affected. If toxin was removed from the diet and proper treatment administered, individuals could recover within two months.

After World War II, alimentary toxic aleukia was eradicated in Russia primarily through education efforts to eliminate the practice of over-wintering grains.

Trichothecene Mycotoxins as Biological Weapons

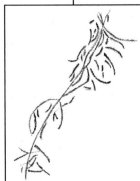

Fusarium

Fusarium molds were first used in Russia to contaminate flour. Shortly after World War II, the flour was used to bake bread for civilian consumption. Some victims developed alimentary toxic aleukia and accompanying gastrointestinal symptoms, fever, chills, muscle pain, depression of bone marrow, and sepsis. Those who survived further into the illness developed painful ulcerations of the pharynx and larynx, bleeding, gastrointestinal ulcers and erosion of the lining of the gut.

During a period from 1979 to 1981, these mycotoxins were purportedly used in biological warfare efforts in Laos, Cambodia and Afghanistan. Distributed in liquid form, they became known as "yellow rain" and caused over 10,000 deaths, primarily among civilians, in these countries.

When these toxins are exposed to the skin, they penetrate within minutes and cause severe, painful burning lesions, tissue death seen as blackening of the skin, and sloughing of perhaps large areas of epidermis. If inhaled, the throat and nose become painful and the victims may have nosebleeds. They will sneeze, cough, wheeze and experience shortness of breath, accompanied by chest pain and spitting up of blood. If the toxin gets into eyes, tearing, pain and blurred vision will occur. A final route of exposure is through ingestion, which results in mouth and throat pain, bloody saliva, nausea, vomiting, abdominal cramps, and diarrhea that may also contain blood. Severe poisoning by inhalation or ingestion will lead to a body-wide condition and lack of muscle control, general weakness, prostration, low blood pressure and dizziness, rapid heartbeat, collapse, shock and death. This progression may take place over minutes or days.

Diagnosis, Treatment and Prevention:
Trichothecene Poisoning
Victims of this type of poisoning will have an elevated white blood cell count. Various assays, including gas chromatography and mass spectrophotometry, can be used to detect toxin in patient blood or urine, or in environmental samples. If the toxins are ingested, activated charcoal can be administered. Otherwise, treatment is limited to supportive care.

As of this writing, there is no antitoxin available for trichothecene mycotoxins, but a vaccine and other pretreatment regimens are being studied in animals. These toxins are very heat stable, requiring thirty minutes at 500°F for inactivation. They are also stable under ultraviolet light, but they can be destroyed with chlorine or exposure to sodium hydroxide. Skin can be decontaminated with soap and water, or a solution of ten percent bleach and three percent hydrogen peroxide.

In case of aerosol delivery, protective masks and clothing could be worn. If toxin were to get into victim's eyes, it could be removed with adequate irrigation using saline solution.

Mycotoxins as Contaminants of Food or Water
Similar to other natural toxins, water would be a poor vehicle because of diluting affects. Food, however, as was proven in Russia, could effectively deliver these toxins with devastating effects.

Natural Toxins—Ricin, Marine Toxins, Mycotoxins

Disease	Symptoms	Transmission	Prevention / Treatment	Food / Water Sabotage
Ingested ricin	Severe GI distress	Ingest castor beans	Activated charcoal	Possible but need large amount / No
Inhaled ricin	Tight chest, cough, fever, nausea, short breath	Inhale aerosols	Supportive therapy	No / No; (aerosol dispersal possible but need large amount)
Histamine poisoning (Scrombroid)	Rash, headache, diarrhea, vomiting	Ingest finfish that are not fresh	Antihistamines, epinephrine	Not likely / Not likely
Ciguatera	GI symptoms, burning, weak, dizzy, itching, myalgia, temp. reversal	Ingest contaminated large reef fish	Induce vomiting; activated charcoal; supportive therapy	Not likely / Not likely
Paralytic shellfish poisoning	Numbness, tingling, dizzy, drowsy, headache, uncoordination	Ingest shellfish associated with red tides	Same as above; respiratory therapy	Possible / Not likely–inactivated by chlorine (aerosol delivery possible)
Neurotoxic shellfish poisoning	Similar to paralytic plus tingling of mouth & throat, myalgia, temp. reversal	Ingest dinoflagellate-contaminated shellfish from Gulf & South Atlantic states	Same as ciguatera	Not likely / Not likely
Amnesic shellfish poisoning	GI & neurological symptoms, short-term memory loss	Ingest shellfish contaminated with toxin-producing diatom	Same as ciguatera	Not likely / Not likely
Aflatoxicosis (Aspergillus)	Vomiting, anorexia, fluid accumulation, jaundice; liver cancer	Ingest contaminated grains, milk or peanuts	Activated charcoal, supportive therapy	Possible / Not likely
Ergotism (Claviceps)	Gangrene; burning sensation; convulsions	Ingest contaminated grains, i.e., rye & wheat	Activated charcoal, supportive therapy	Possible / Not likely
Alimentary toxic aleukia (T2)	GI inflammation, vomiting, diarrhea, skin hemorrhage	Ingest contaminated cereal grains	Activated charcoal, supportive therapy	Possible / Not likely (aerosol or "yellow rain" delivery possible)

CHAPTER ELEVEN

Chemicals and Pesticides

The concerns of the Food and Drug Administration and other organizations regarding biological warfare aimed at our food and water supplies have involved primarily biological agents. There are, however, several toxic agents that are chemical in nature that could be used to contaminate food or water with devastating affects on human health. Chemicals may in fact be more appealing to terrorists than biological agents because they mean a decreased risk to the perpetrator. Many chemical agents are readily available, and the handling of such agents requires little or no expertise.

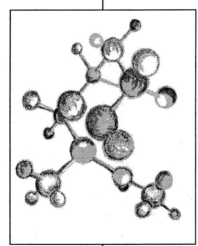

Chemical model

Some chemicals are harmless by themselves, but become dangerous when used in combination with other chemicals. This is true of certain cleaning agents such as chlorine bleach and ammonia that, when mixed together, produce toxic fumes. An example of this happening in foods occurred in the past when cobalt was added to beer to improve the quality of the foam; the combination of cobalt and beer yielded a toxic compound capable of destroying heart muscle tissue.

155

Terrorist Use of Chemicals in History

"Mustard gas" is an inclusive term for several man-made chemicals that occur not as a gas at room temperature, but rather as colorless, odorless liquids. If these are mixed with other chemicals, they take on a brownish color and tend to smell like garlic. During World War I, Germans utilized mustard gas. It was again introduced in World War II, when large amounts were made and used by both the Allied forces and the Germans, causing burns to the skin and when inhaled, to lungs. Mustard gases were also used in Iran and Iraq from 1984 to 1988. In the United States today, mustard gas is used only for research.

The nerve agent sarin is a true, highly volatile gas. It is odorless, tasteless, and colorless, and more deadly than cyanide. Sarin gas was released by terrorists into a subway system in Tokyo in 1995, killing twelve and injuring over 5,000, though seventy-five percent of the injuries were classified as minor.

In a totally different vein, common fertilizer was the weapon used by Timothy McVeigh to blow up the Federal Building in Oklahoma City in that same year.

Chemicals & Poison Gases

There are literally hundreds of chemicals and poisonous gases in existence today, most of which have legitimate uses in industry. New chemicals are being developed and introduced every year. The Centers for Disease Control have eight categories of chemical agents, based primarily on how these substances affect the human body. The list below includes these categories and provides an example of a chemical for each.

- Nerve agents (sarin)
- Blood agents (hydrogen cyanide)
- Blister agents (mustard)
- Pulmonary/choking agents (chlorine, phosgene)
- Incapacitating agents (LSD)
- Poisonous industrial chemicals (organic acids)
- Riot control agents (tear gas)
- Vomiting agents (adamsite)

Additional health threats include heavy metals such as arsenic and mercury.

Sarin: A Nerve Gas

Sarin is a poison gas that affects the nervous system, interfering with the nerve impulses that travel from the central nervous system to the various organs of the body. Upon exposure to this gas, victims initially experience eye symptoms including pain, redness, small pupils and blurred vision. If the skin comes into contact with sarin, muscles near the skin's surface will actually start to twitch when they become over-stimulated. Initial exposure also causes a runny nose, headaches, nausea and vomiting, as well as excessive salivation. When the agent is inhaled, the victim will cough and be short of breath as airways become constricted. If the dose is high, victims may lose control of their bowel and bladder. There may be a sudden loss of consciousness, or the victim may convulse, perhaps ceasing to breathe altogether. Death may occur within one to ten minutes after inhaling sarin.

General symptoms of exposure to nerve gas, which ensue very rapidly following exposure, include the aforementioned involuntary muscle twitching, small pupils, and shortness of breath. Hazardous material "hazmat" teams have special equipment that will detect the presence of nerve gas and other chemical agents. There are also biochemical tests available to assay victim's blood. After the nerve agent is identified, antidotes, if available, can be administered. These antidotes work by blocking the effects of the chemical. Other therapy is supportive, such as the use of oxygen or anticonvulsants.

Obviously the best response to nerve gas exposure is to leave the area immediately and breathe fresh air. Nerve agents are actually heavier than air and will sink toward the ground, so going to upper floors of a building would offer some degree of protection. If eyes are exposed, they should be flushed with warm water for at least fifteen minutes. Exposed skin should be thoroughly washed with soap and water followed by a rinse with dilute chlorine solution, but contaminated clothing must be bagged and incinerated. If the victim's hair becomes contaminated, it must be shaved, bagged and burned as well. Activated charcoal can be administered to alert victims, but vomiting should not be induced.

Those members of a hazmat or other responding clean-up team should wear protective gear including suits, masks and gloves. They should not administer cardiopulmonary resuscitation to victims whose facial skin has been exposed to the nerve agent.

Hydrogen Cyanide: A Blood Poison

Hydrogen cyanide is a colorless gas or liquid that is an eye, skin and respiratory tract irritant, as well as a blood poison. When inhaled, this chemical interferes with the central nervous system and causes drowsiness, confusion and shortness of breath. If enough is inhaled, hydrogen cyanide will cause death.

Mustard Gas: A Blister Agent

Mustard gas is an eye irritant as well as a blister agent. It causes burns and blisters on the surface of the skin. If inhaled, the victim will initially experience coughing, however, such exposure can lead to bronchitis, other long-term respiratory problems or even lung cancer. If the initial dose is large, mustard gas can be fatal immediately. There is no antidote.

Chlorine and Phosgene: Pulmonary Agents

Pure chlorine is a greenish-yellow liquid that has a strong, pungent odor. It is very corrosive to the skin and eyes, and if it comes into contact with these areas of the body, can cause blurred vision and burns. Chlorine easily becomes a gas, and when it is inhaled, the victim will have labored breathing due to fluid build-up in the lungs. A high amount of inhaled chlorine gas will kill immediately.

Phosgene is a colorless gas that is used in the production of certain chemicals. If high concentrations are inhaled over time, as might be the case for workers in facilities that use it, phosgene can cause severe breathing difficulties and perhaps fatal congestion of the lungs. There is no known treatment for exposure to phosgene, and the mortality rate is high, though some people exposed to significant levels do survive.

Chemicals Commonly Found in Foods

Chemicals such as pesticides routinely end up in our food, and many chemicals, like preservatives, are intentionally added for the purpose of shelf life extension. Chemical food poisoning is extremely rare. It is a form of intoxication that is usually due to a chemical or pesticide that accidentally gets into the food in high enough amounts to cause illness. When it occurs, food poisoning due to ingestion of chemicals is characterized by a very short onset time.

Pesticides

The Environmental Protection Agency defines a pesticide as any substance intended to destroy, prevent or repel pests such as insects, weeds, fungi and rodents. Pesticides are commonly used in the United States to increase crop yields, thereby minimizing cost to the consumer. Certain types of crop pests can also produce toxins that are harmful to humans when they are ingested so preventing growth of these pests is imperative. Pesticide use has enabled farmers to grow a larger variety of crops, which has benefited all Americans.

That is the upside. Unfortunately, pesticide use has also harmed the environment. Many pesticides used in the past, such as DDT, have been banned after they were discovered to be toxic to animals and/or humans, or found to accumulate in the ecosystem.

Examples of Pesticides

Hexachlorobenzene

Grains are subject to contamination with chemicals. Hexachlorobenzene was a commonly used pesticide until 1965, when it was discovered that it accumulated in the environment, where it also persisted over time and became concentrated as it moved up the food chain. Humans became contaminated through consumption of meat, poultry or dairy products if food animals had eaten tainted feed. Affects of hexachlorobenzene in humans include abnormal fetal development and lowered survival rate of young children. Adult symptoms involved the kidneys, blood cells, bones and the immune system.

DDT

The Environmental Protection Agency was formed in 1970, and one of the very first actions it took, after being sued by citizens groups, was the banning of DDT, which was the most powerful pesticide in use at the time. DDT is able to destroy a myriad of pests, including many species of insects, and was used to control malaria-carrying mosquitoes during World War II. This potent substance accumulates in the fatty tissue of animals and humans, where it can lead to cancer and genetic abnormalities.

Chlordane

The insecticide chlordane, originally used in 1948, was banned by the EPA in 1988 due to its tendency to become concentrated in the environment (bioconcentration). Once it is absorbed, chlordane has a tendency to persist in an aquatic environment as well. It may still be present in dangerous amounts in the tissue of fish.

Toxaphene

Toxaphene is the trade name of another pesticide that is actually a combination of over 670 chlorinated camphenes. This pesticide, which can persist in the environment similar to hexachlorobenzene, gets into surface waters as a result of runoff. In humans, toxaphene is stored in fatty tissues like DDT. It affects the central nervous system and the liver, and the EPA has flagged it as a possible carcinogen. Those particularly vulnerable to the affects of toxaphene are the typically susceptible populations: pregnant women, fetuses, nursing mothers and their infants, and children.

Controlling & Preventing Misuse of Pesticides

Several government agencies and organizations work together to monitor the use of pesticides. Pesticide residues on foods are subject to *action levels* established by the FDA and the EPA. If a food has residues of a pesticide that exceed these levels, the product cannot be sold. Some pesticides subject to action levels, such as DDT, may still persist in the environment even though they are no longer used, or may be found in imported foods.

The EPA is responsible for reviewing pertinent scientific data as it becomes available and before new pesticides are registered for use. A tolerance level must also be determined for pesticides that will be used on human foods. The amount of pesticide residue allowable in foods is strictly regulated, and presence of trace amounts does not mean the food is unsafe. Pesticide manufacturers must conduct extensive toxicity tests on new products before they can be released for use on food that is directly or indirectly destined for human consumption. These tests, which are sensitive enough to detect the equivalent of one teaspoon of salt in one million gallons of water, determine what is called the "No Toxic Effect Level," or NOEL. This is the maximum level of residual pesticide on foods at which there are supposedly no adverse (toxic) effects.

These tests are designed to determine these levels in the most susceptible populations—children and the elderly. The pesticide residue levels that are legally allowable in foods are then set to between ten and one hundred times *lower* than this established NOEL. This safety factor ensures that even if someone eats an extraordinarily large amount of a certain food containing pesticide residue, or different foods that have all been treated with the same pesticide, ingested levels will remain below the NOEL.

The Food and Drug Administration, the Environmental Protection Agency, and the Food Safety and Inspection Service branch of the United States Department of Agriculture work together to enforce NOEL and action levels. The FDA assays certain foods for pesticides as close as possible to the point of production. If the NOEL is exceeded, the food will be subject to regulatory action such as a seizure or injunction. If the food is an import, the FDA can stop the shipment at the port of entry. Foods that are not under FDA jurisdiction—meat, poultry, and some egg products—are monitored by the United States Department of Agriculture.

Fortunately, studies indicate that washing of fresh produce will eliminate many pesticide residues. This can be done using copious amounts of warm or cool water and a scrub brush, if appropriate, but no soap. Removal of the outer leaves of lettuce and cabbage will also remove most pesticide residues. Since pesticides tend to accumulate in fatty tissue, trimming of fat and skin from meat, fish and poultry will help reduce pesticide levels in these foods. Eating organic foods will also help to reduce susceptibility to chemical poisoning through pesticides.

Certified Organic Foods

Starting in October 2002, the United States Department of Agriculture began requiring that anyone making or selling certified organic foods label these products as such. "Certified organic" means that no growth hormones or antibiotics were given to food animals, and that most pesticides and chemical fertilizers have not be administered. The restrictions also preclude the use of ionizing radiation, sewage sludge and bio-engineered products. Products bearing the label "100 percent organic" must contain only organic ingredients, and those labeled simply "organic" must contain a minimum of ninety-five percent organic ingredients. Organic foods range from meat, fruits and vegetables to cheese, bread and potato chips.

Other Chemicals Found in Foods

Arsenic—A Heavy Metal

Arsenic is an element, but it usually is found in nature in combination with another element, like oxygen or carbon. Such compounds are white or colorless powders with no smell or taste, so they are undetectable in food, air or water. Pure arsenic is imported into the U.S., and is used as a preservative in pressure-treated wood. Arsenic combined with carbon or hydrogen in organic compounds is still used today as a pesticide. Arsenic metals are used in the manufacture of lead-acid batteries, semiconductors and light-emitting diodes.

In the environment, people are exposed to arsenic through food, air and water, or by skin contact with soil or water containing arsenic. Fish and seafood contain the highest levels of arsenic as an organic compound, which is the less harmful form. Inorganic arsenic, which is arsenic combined with elements such as oxygen or chlorine, has long been recognized as a poison. Ingestion of large doses (over 60,000 parts per billion) can result in death. At 300 to 30,000 ppb, the victim will experience vomiting, diarrhea, stomachache, possibly a decrease in production of blood cells, which causes fatigue and cardiac abnormalities such as blood vessel damage and an abnormal heart rhythm.

Long-term ingestion of small amounts of arsenic affects the skin, causing darkening and the formation of small warts on the palms, soles and torso, some of which may become cancerous. Small amounts of ingested arsenic have also been linked to cancers of the liver, gall bladder, kidneys, prostate, and lungs; inorganic arsenic is a known carcinogen. Breathing high levels of arsenic leads to throat and lung irritation, perhaps skin manifestations, as well as circulatory and peripheral nerve damage.

Mercury

Fish are subject to contamination with mercury, both from natural accumulation due to acid rain, and from industrial processes. Mercury becomes concentrated as it is passed along the food chain, so fish that contain the highest levels are those toward the top of the chain, such as salmon and tuna. Pregnant women are particularly susceptible to mercury poisoning from eating fish. This type of poisoning results in low infant birth weights, and lowered infant IQ accompanied by below-normal behavioral assessments. Nursing mothers and their infants, as well as children, are also more susceptible than the general population to mercury poisoning. Fish may also contain PCBs, chlordane, dioxins and DDT.

PCBs

Polychlorinated biphenyls (PCBs), the by-products of industrial processes like paper manufacturing, are found in the soil and river sediments. Overexposure to PCBs can produce a skin rash, damage the liver, gastrointestinal tract and the blood, and affect the immune, reproductive and endocrine systems. PCBs also become concentrated in breast milk. Vegetables that grow underground, like potatoes and carrots, are particularly susceptible to these chemicals, but any type of food grown in the soil can be affected since PCBs are incorporated into the plants as they grow. The plants are then consumed directly by humans, or are consumed by food animals that are eventually eaten by humans. These chemicals are difficult to detect in foods, so the best strategy is to eat a variety of different types of fruits and vegetables from different sources, wash them thoroughly with warm water prior to consumption, or buy organic produce. (Be aware that organic produce is fertilized with animal or perhaps even human waste, and carries risks of its own.)

Dioxins

Dioxins are by-products of incineration and combustion processes, and the uncontrolled burning of residential and landfill waste, which is currently believed to be the greatest source of dioxins. Dioxins are also produced when chlorine is used as a whitener in the paper and pulp industry, and they are present as contaminants in certain kinds of chlorinated organic compounds, though efforts of the EPA have greatly reduced these and other industrial sources.

The EPA estimates that as much as ninety-five percent of dioxin intake by humans comes from ingestion of animal fats, while minor sources include air and water. Dioxins find their way into foods via the air, which contaminates the soil, water, and plants. Certain levels of dioxins are found in various types of foods, especially animal fats, milk and other dairy products. Similar to PCBs, dioxins have detrimental affects on the immune system, nervous system, reproductive system, endocrine system, liver, gastrointestinal tract, skin and blood.

163

Cleaning Chemicals

Certain disinfectants used to clean equipment in food processing operations are also detrimental to human health. Chlorhexidine diacetate is used to clean surfaces in meat, poultry and egg processing facilities. Although proper use of this compound followed by sufficient rinsing should prevent this chemical from contaminating foods, residues left on surfaces will leach into food and can result in eye and liver damage, especially among women.

Propionic Acid

Propionic acid is used to control the growth of bacteria and molds in stored grains and drinking water used for livestock and poultry. It is also commonly used to prevent mold growth on commercial bread products. Humans are exposed to propionic acid when they eat food crops, bread, milk, poultry, or meat. While such exposure is not normally dangerous, it may cause ulcers. Direct contact with concentrated propionic acid may damage skin, eyes and mucous membranes. It is also somewhat toxic if inhaled.

Acrylamide

The Food and Drug Administration has recently begun researching the presence of acrylamide in fried and baked starchy foods such as potato chips and french fries. The amino acid asparagine which occurs naturally in these foods, when heated to high temperatures in the presence of certain sugars leads to the formation of acrylamide. Though it is only suspected of causing cancer in humans, it has been shown to do so in laboratory animals. Scientists in Sweden have accused acrylamide of perhaps being responsible for several hundred cases of cancer per year in their country.

Chemicals and Pesticides as Potential Weapons

Sarin and mustard gas have both been successfully used as weapons in the past, and the potential is there for future use as well. Virtually any chemical could be used in an attempt to contaminate food or water, but mustard would not be a good choice since it does not easily dissolve in water. It also breaks down quickly in water and in soil, and would presumably do so in foods as well. It does not build up in animal tissue.

Perhaps a more logical route would be the use of pesticides, since they are easier to obtain than nerve gas and blister agents. In the past, the Environmental Protection Agency has focused efforts on the safety of pesticide manufacturing plants, their operation, equipment and training of employees. Distributors have been concerned with proper storage and labeling of pesticides, and end users have been responsible for reading and following instructions. Currently, however, additions are being made to the list of precautionary measures.

The EPA has made several recommendations to improve the security of pesticides. These include increased security measures for buildings, manufacturing and storage facilities, and surrounding properties through the use of fencing, lighting, locks, intrusion detection systems, cameras and even guards. Equipment that is used to transport and apply pesticides should be secured, and drivers and handlers of such equipment should have proper training and identification.

The Federal Bureau of Investigation has requested that aircraft used for pesticide applications be under vigilant security, and that handlers and pilots be particularly aware of suspicious activity. This may include the undesignated use of an aircraft, a request for training in the use of aircraft or dangerous chemicals, or attempts to acquire such chemicals. Threats, unusual purchases, suspicious behavior by employees or customers, and unusual contacts with the public should also fall under severe scrutiny.

In addition to protection of chemicals and equipment, the EPA suggests guarding confidential information about pesticides that may be obtainable via computer or communications systems. Facilities and equipment should be designed to minimize the risk of damage by an intruder such as a saboteur or a computer hacker. Procedures and policies should be developed regarding the hiring and training of employees, and should include background checks of those who will be authorized to access restricted areas, such as pesticide storage. It is also extremely important to monitor inventory, and to have an emergency response plan in place in case of a break-in, other terrorist activity, or an accident. Authorities (local police, FBI) must be notified if there are any reports of pesticide use or exposures that are unusual and outside the realm of conventional patterns.

Intentional Contamination of Food or Water

Unfortunately, chemical contaminants may not be noticeable in a food. The FDA indicates that food tampering is possible at several points before the food goes to wholesalers or end consumers, and that prevention and response capabilities are, at this point, limited. To be effective, a terrorist would have to contaminate a food that is consumed regularly and in large quantities, such as infant formula or orange juice. The container itself would have to be easily subject to tampering. Water supplies would be less likely to be used as vehicles for chemical poisoning due to diluting effects.

There are tools available, some of which are portable, to detect most of the toxic elements that would be practical for terrorist use. These can be used after the fact, but the problem of prevention remains. It is imperative that the supplies of chemicals and pesticides be scrupulously monitored and guarded so that they do not fall into the wrong hands.

Chemicals and Pesticides

Agent	Symptoms	Transmission	Prevention / Treatment	Food / Water Sabotage
Nerve gas (i.e., sarin)	Muscle twitches, eye symptoms, runny nose, nausea, vomiting, cough, shortness of breath	Inhalation or skin contact	Some antidotes available / Supportive therapy	No / No (aerosol delivery possible)
Mustard gas	Skin burns and blisters; eye irritation, cough	Inhalation or skin contact	Supportive therapy	Same as above
Pulmonary agents (chlorine, phosgene)	Blurred vision, skin burns; labored breathing if inhaled	Chlorine—inhale gas; contact skin with liquid Phosgene—inhale	Supportive therapy	Same as above
Hydrogen cyanide	Drowsiness, confusion, shortness of breath; skin & eye irritation	Contact skin/eyes with liquid or inhale gas	Supportive therapy	Possible / Not likely (aerosol delivery possible)
Arsenic	Ingestion: warts on palms & soles; cancer	Ingest small amount in food and water	Highest levels found in fish and seafood	Possible / Possible
Pesticides	Various, including cancer	Same as above	Wash fruits & vegetables; trim fat on meats; remove outer leaves	Possible / Possible
Mercury	Birth defects; brain, liver, kidney poisoning	Ingest fish from contaminated water	Avoid contaminated fish	Not likely / Not likely
PCBs	Rash; systemic damage	Vegetables grown in contaminated soil	Wash fruits and vegetables; eat a variety; buy organic produce	No / No
Dioxins	Systemic damage	Ingestion of milk, dairy, animal fats	Eat a variety of foods; trim fat	No / No
Chlorhexidine (cleaning)	Eye and liver damage	Ingest foods contaminated with residues (meat, poultry, eggs)	Eat a variety of foods	Possible / Possible

PART THREE

Livestock
and Crops

CHAPTER TWELVE

Bovine Spongiform Encephalopathy & Creutzfeldt-Jakob Disease

This chapter addresses yet another type of disease agent called a prion. Prions are so recently discovered that their mode of action is not yet fully understood. The two illness discussed below are types of *transmissible spongiform encephalopathies* (TSEs) that are

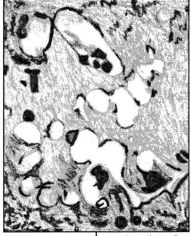

Kuru-infected human brain tissue

caused by prions. *Bovine spongiform encephalopathy* (BSE) in cows, and its human counterpart, *Creutzfeldt-Jakob Disease* (CJD), are both degenerative diseases that cause brain tissue to become holey like a sponge, therefore the term "spongiform."

TSEs represent an atypical type of food-borne illness, but one that produces a fatal disease of the central nervous system. As of yet, BSE and CJD are not recognized as having the potential for use as biological weapons, but these diseases, along with other forms of TSEs, are mysterious, rare, and always fatal. Most importantly, they can be acquired though consumption of contaminated foods.

The History of BSE & CJD

The first historical evidence of human transmissible spongiform encephalopathy was a disease known as "kuru." It was common among members of a native tribe in the highlands of Papua-New Guinea, who referred to the illness as "laughing death." Prior to the late 1950s, tribal members partook in a death ritual that included eating body parts of the deceased, even the brain. Male tribesmen were given the choice muscle tissue, but children and women, who developed kuru more frequently, ate the brain and other less desirable offerings. Dr. D. Carlton Gajdusek originally made the connection between these obscure death rituals and kuru. He informed the tribes of his findings, and the morose practice was abandoned. Dr. Gajdusek went on to win the Nobel Prize for his work on kuru and CJD.

The original form of Creutzfeldt-Jakob Disease was identified in 1920 by two German scientists, Creutzfeldt and Jakob. Classic CJD is a disease of older people who will suffer only two to six months before succumbing to this illness, usually around age sixty-eight or over. Similar to bovine spongiform encephalopathy, CJD in humans causes symptoms that are consistent with the gradual destruction of brain tissue, which may start with simple things like forgetfulness or absentmindedness, and progress to other psychiatric, sensory and neurological problems, ending with dementia and death. While CJD cannot be conclusively diagnosed until the time of autopsy, a brain X-ray or magnetic resonance imaging (MRI) taken in the later stages of the illness will show evidence of tissue deterioration.

Cases of CJD began to appear in young people in the 1980s who were given growth hormone derived from the pituitary glands of diseased cadavers. This practice continued in several countries, including the U.S., until 1985 when a switch was made to synthetic growth hormone. This illness has also been contracted via cornea transplants, and has been transmitted in the operating room by improperly disinfected instruments.

An unexplained increase in the incidence of CJD began in 1989. Not only was the rise odd, but once again, young people were being affected with what was typically an illness of the elderly, one that caused neurological problems and impaired speech, vision, thought processes, and even motor skills. That same year, the National Institutes of Health held an international workshop on BSE with presenters from the UK, Italy, Germany, France and the United States.

Bovine spongiform encephalopathy (BSE) is a highly transmissible TSE among bovines. Almost all of the known cases of BSE (ninety-five percent) have occurred in the United Kingdom, though a few have been reported from France, Belgium, Spain, Switzerland and Germany. There have been no official cases of BSE in the United States to date. BSE is a chronic and degenerative illness of the central nervous system for which there is no treatment and no cure. Veterinarian Dr. Colin Whitaker first described BSE in Great Britain in 1986 as a "new, scrapie-like syndrome," scrapie being the same type of illness in sheep. The symptoms of BSE are very similar to those of scrapie. Cows become anxiety-stricken, stop eating, lose their sense of balance and direction, and eventually waste away and die. "Mad Cow Disease" garnered a lot of attention when, by May of 1987, it had reached epidemic proportions in Great Britain. It did not reach its peak until 1993 when there were over 1,000 new cases being reported every week. When humans ate infected beef, they could develop a new form of Creutzfeldt-Jakob Disease.

The CJD associated with contaminated beef is slightly different than classic CJD, and is called "variant CJD" or "vCJD." It affects predominantly younger people, who will live with the disease for six to twenty-two months, dying at the average age of twenty-eight. Because of the long incubation time (ten to twenty years), victims become infected when they are only children. The first identified victim of vCJD was a British eighteen-year-old who died in May, 1995. Classic CJD is endemic worldwide, including the U.S., but vCJD, so far, is limited to Great Britain and a few other European countries. One young woman in Florida was diagnosed with vCJD, but she was formerly a citizen of the UK and had protracted the illness there. As of July 2002, there have been 125 cases of vCJD, 117 in the UK, six in France, and one each in Ireland and Italy. Most of these victims had eaten infected beef over several years, from 1980-1996, during the peak incidence of BSE in Great Britain. It is estimated that since the beginning of the BSE epidemic in England, at least 750,000 infected cattle have entered the human food chain.

Although the link between BSE and vCJD had been suspected for a long time by agricultural and medical researchers, in March of 1996, the British Spongiform Encephalopathy Advisory Committee informed the British government that vCJD was quite likely to be caused by eating or coming into contact with beef infected with BSE. From the onset of the epidemic in 1987, the number of BSE cases in England had increased from a handful to over 150,000. As a result of this disclosure, beef sales plummeted as many countries boycotted beef from the UK.

How Were Cows Originally Infected with BSE?
As was discerned by Dr. Whitaker in 1986, BSE is very similar to scrapie in sheep. While BSE is restricted to cows in the UK, scrapie is found worldwide, as was previously the practice of using rendered, diseased animals as protein sources in animal feed. Scrapie is so named because infected animals scrape up against fences, as if scratching themselves, to the point where the wool is scraped from their hide. Other symptoms are loss of appetite, balance and sense of direction. Animals become nervous and apprehensive, and eventually waste away. Death is inevitable. Scrapie has been endemic in sheep in England for over two hundred years, but it first appeared in the U.S. in 1947, apparently imported from England along with sheep. Every U.S. state save eleven had reported scrapie by 1990.

The epidemic of BSE and vCJD in Great Britain erupted as a result of conservation efforts. Sheep that had died of scrapie were not burned or buried, but rather were recycled into a protein supplement used in the production of animal feeds. Cows are herbivores (plant-eaters), but over the years, farmers have found that supplementing cattle feed with extra protein increases yields of milk and beef. Thus began the practice of adding ground-up animal protein to cattle feed. This protein included meat-and-bone meal from scrapie-infected sheep. Calves were even fed rendered bovine meat-and-bone meal, thereby becoming not only omnivores but cannibals as well.

Apparently BSE can also be passed from a mother cow to her offspring, as demonstrated in studies of calves that were never fed protein supplements, yet still developed BSE. There is also evidence that this has happened in humans. The New England Journal of Medicine reported that a thirty-nine-year-old woman developed symptoms of CJD in her thirtieth week of pregnancy. She gave birth to a healthy baby, but placental and umbilical cord material injected into mice resulted in spongiform illness in the mice.

The Gruesome Truth About Rendering

When a cow goes to slaughter, nothing goes to waste. Only about fifty percent of a cow's weight is in the muscle tissue. The intestines and their contents, the head, hooves, horns, bones and blood comprise the rest, all of which goes from the slaughtering floor and into huge grinders. Whole animals are sometimes added as well—food animals that are diseased and cannot be used for meat, but also six to seven million euthanized pets per year and even roadkill. The disgusting mix is heated, the fat is extracted, and the rest is turned into "protein" powder. About forty billion pounds of dead animals are processed each year in the $2.4 billion a year industry known as rendering. One of the by-products of beef production is gelatin, a conglomeration of hooves, hides and spinal columns that is used in manufacture of pharmaceutical capsules, candy, yogurt and even cosmetics.

In the late 1970s, a few changes were implemented that may have exacerbated the BSE problem in England. The use of strong solvents to extract tallow and fat was abandoned, and systems were changed from batch operations to more efficient, continuous processes that allow contaminants to be distributed among lots. Cooking temperatures were also reduced to enhance flavor and decrease energy costs. In this way, scrapie-infected sheep were ultimately fed to cattle—only one teaspoon of highly infective feed concentrate can cause BSE in a cow.

While there have been no cases of BSE reported in the U.S., about 100,000 cows die each year in the United States from unknown causes that are lumped together under the heading "downer cow syndrome." These downer cows go directly into rendering, and it is possible that at least some of these cows may have been infected with a BSE-like agent. By late 1997, the FDA had placed a ban on the feeding of ruminant protein to ruminants, but protein from horses, pigs, chickens, dogs, cats, and turkeys, as well as fecal material and blood from chickens and cows, are still used.

Prions: A New Agent of Disease

In the late 1990s, no one had a clue as to what the agent of BSE and CJD was. They knew it was not a bacterium or a virus because it could withstand intense heat, radiation, and exposure to most disinfectants. In 1982, Dr. Stanley B. Pruiner of the University of California-San Francisco suggested that scrapie was caused by "prions." Prions are defined as self-replicating, aberrant protein molecules that are a modified form of a normal component on cell surfaces. Prions are pathogenic forms of these proteins that possess the ability to induce changes in normal brain proteins, though the means by which this occurs is not yet fully understood.

What is known about prions was derived from animal studies. Animals were either fed material from TSE-infected animals, or this material was injected directly into the brain. Test animals developed symptoms of TSE, especially when the material came from the brain, thymus, tonsils, and spinal cords of the TSE-infected animals. These parts are not normally eaten as such, but can end up in processed meats such as hamburger, sausage, meat pies and other products. The theory that prions are the disease agents of BSE and CJD is widely accepted, though some scientists think there may be another contributing factor, perhaps a virus.

Other Types of TSEs

Several animal species suffer from forms of TSE. Goats as well as sheep can get scrapie; mink and cats can also be affected by spongiform encephalopathies specific to those species of animals, and deer and elk share a TSE known as *chronic wasting disease*, or CWD.

CWD originally was a problem in western states. It was first detected as a disease in 1967 in Colorado, and recognized as a form of TSE in 1978. It now has spread eastward, necessitating the destruction of thousands of deer in the state of Wisconsin in the summer of 2002. It has also reached Kansas, Oklahoma, Nebraska, South Dakota, Wyoming, Montana and Alberta, Canada.

So far, no evidence exists that CWD can affect people, but the deaths of three men during the 1990s in north central Wisconsin are under investigation. These men were friends and hunting companions who ate elk and deer meat on a regular basis during the 1980s and 1990s. Although there has never been a confirmed case of human illness attributed to consumption of wild game, officials advise hunters not to eat the parts of deer most susceptible to contamination, such as the brain and spinal column, and to not eat any part of a deer that appeared to be sick. The CDC also contend that it may be *possible* to contract CJD in this manner. The Wisconsin Department of Natural Resources also states that CWD is not likely to be transmitted to livestock.

What is Being Done to Control TSE?

Starting in 1989, the U.S. government placed severe restrictions on the importation of live cattle, sheep, and goats, and on certain ruminant products, from all countries with known cases of BSE. In 1997 the ban was extended to include ruminants and ruminant products from all European countries. This ban includes all meat products, including those used in animal and pet foods. Milk and milk products are not believed to be cause for concern, so their importation into the U.S. is still allowed.

Importation of products that contain gelatin, such as capsules or certain kinds of candy, is still allowed, but the FDA requires that specific guidelines be followed so that these products will be safe. These guidelines include the use of only BSE-free cattle in the manufacture of such products. Cosmetics and dietary ingredients containing bovine material from the thirty-one countries identified as being at risk for BSE, (see below), are also prohibited from import into the U.S.

In 1990, the USDA began a program of active surveillance for BSE in the United States. By late 2000, the USDA banned all imports of rendered animal protein from the thirty-one designated countries. The FDA no longer allows production or use of cattle feeds containing protein from most mammals. The FDA also began an education and compliance program for renderers, feed mills, dairy farms, distributors, protein blenders and even feed haulers. The CDC has enhanced its surveillance of vCJD as well, and has found no cases originating in the U.S., nor has any BSE been found. However, according the FDA, BSE and similar diseases are very difficult to study due to long incubation periods, and though the risk is low that BSE could gain entry to the United States, it may still be possible.

If You Travel

Thirty-one countries, mostly in Europe, have been identified as having or being at risk for BSE. They are: Albania, Austria, Belgium, Bosnia-Herzegovina, Bulgaria, Croatia, Czech Republic, Denmark, Federal Republic of Yugoslavia, Finland, France, Germany, Greece, Hungary, Ireland, Italy, Liechtenstein, Luxembourg, the Former Yugoslavia republic of Macedonia, the Netherlands, Norway, Oman, Poland, Portugal, Romania, the Slovak Republic, Slovenia, Spain, Sweden, Switzerland, and the United Kingdom.

If you are traveling to any of these countries and are concerned about contracting vCJD, simply avoid eating beef and beef products. Select cuts of beef are acceptable, such as steaks, but avoid ground beef and sausages. Pure muscle tissue is unlikely to be contaminated with prions, which are harbored primarily in the brain, spinal column, thyroid, tonsils, etc. of an infected animal. Even in the U.K., according to the FDA, the risk of getting a contaminated piece of beef in only one per ten billion servings. Risks for specific countries cannot be determined due to distribution of product.

Could Contaminated Beef be Used as a Weapon?

BSE and vCJD are very frightening diseases that consistently kill their victims. In order for someone to be infected with vCJD, however, it would be necessary to eat a significant amount of contaminated beef over a prolonged period of time. Therefore, it is not logical, and only remotely possible, that a terrorist would consider the use of these prion-associated illnesses. The long incubation period would also undoubtedly be a deterrent.

It is still important for people to be aware of TSEs, their sources, manifestations, and methods of prevention. Although BSE has yet to be found in the U.S., the current CWD problem has become prevalent in a large number of states, and for those who regularly consume elk or venison, it is definitely cause for concern.

Transmissible Spongiform Encephalopathies

Disease	Symptoms	Transmission	Prevention / Treatment	Food / Water Sabotage
Bovine spongiform encephalopathy (BSE)	In cows: anxiety, anorexia, lose balance and sense of direction	Feed supplemented with contaminated animal protein	Do not use contaminated feed/none	No / No
Variant Creutzfeldt-Jakob Disease (vCJD)	In humans: forgetfulness and absent-mindedness progressing to dementia	Ingesting beef/beef products from cows suffering from BSE	Do not eat products from BSE cows / Supportive therapy, no cure	Not likely / No
Chronic wasting disease (CWD)	In deer and elk: similar to BSE in cows	Unknown	Control animal ranges & populations to prevent spread	No / No

CHAPTER THIRTEEN

Agri-Terrorism

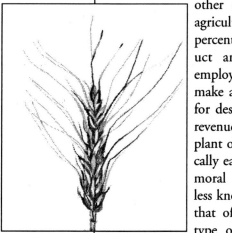

Wheat

According to experts on biological warfare, the purposeful destruction of livestock or crops, "agri-terrorism," may be a very viable alternative to violent terrorism. The United States is particularly vulnerable since it produces more food than any other country in the world. U.S. agriculture is responsible for thirteen percent of the gross domestic product and seven percent of total employment, so a terrorist wishing to make an economic impact may opt for destroying this critical source of revenue. Also, the weaponization of plant or animal pathogens is technically easier, has far fewer ethical and moral complications, and requires less knowledge of microbiology than that of human pathogens. Yet this type of terrorism would devastate both the food production and processing industries, generate loss of confidence in the ability of the government to guard the country against terrorism in general, and bring great financial harm to the United Sates by a reduction in or elimination of food exports.

Historical Aspects of Agri-Terrorism

Although there have been only a limited number of instances in which this type of terrorism was deployed, many countries, the United States among them, have developed plant and animal pathogens for such purposes. During World War I, German saboteurs purposely gave horses sugar cubes tainted with the agents of glanders and anthrax prior to their being shipped to Europe. Since there are very few historical accounts of this incident, it is assumed that the attempt was not particularly successful.

Germany continued their research with agricultural pathogens in World War II, and was joined by Japan, Britain, the United States, and the former Soviet Union. Animal diseases of interest included foot-and-mouth disease, a cattle disease called rinderpest, glanders, and anthrax. Potato beetles and various potato fungi, turnip pathogens and even insects were considered for destruction of crops. Also during World War II, the U.S. developed a virus for use specifically against Japanese rice, and conducted research on avian, bovine, potato, wheat, and other rice pathogens. None of these agents was ever used.

Decades later, Iraq became involved in experimentation with agents of agricultural terrorism prior to the Persian Gulf War. Their potential weapons included wheat stem rust and camel pox, a relative of smallpox. But by far the most extensive agricultural weapons program was undertaken by the Soviet Union, which developed a vast number of plant and animal pathogens for use against cattle, swine, sheep and poultry, wheat, barley, potatoes and tobacco. Because of all the experimentation done in the past, there is an extensive knowledge base regarding the use of these various agents.

Animal Pathogens

The United States has inadvertently increased the vulnerability of livestock to animal pathogens through practices of modern-day husbandry. Farmers have routinely isolated their animals in order to protect them from disease, but this has only made livestock more vulnerable. Other methodologies have added to this vulnerability as well. Animals are under a great deal of stress due to crowded conditions, which lowers their resistance. The administration of antibiotics, steroids, and other drugs used to increase the volume and quality of milk and meat production also makes animals more vulnerable to viral and bacterial infections. A pathogen introduced to an animal herd thus tends to spread quickly, and is often passed directly from mother to offspring during gestation. Finally, with the eradication of many animal pathogens such as anthrax and foot-and- mouth disease, vaccination programs have ceased, making these and other pathogens possible agents of this type of terrorism. Bovine spongiform encephalopathy and anthrax have been discussed in previous chapters. Other animal pathogens of major concern are discussed here.

Foot-and-Mouth Disease

Foot-and-mouth disease (FMD) is caused by a virus that is capable of infecting all types of cloven-hoofed animals (those having hooves that are divided into two parts), as well as some other animals. It causes a severe illness in cattle and swine, but sheep, goats, deer, hedgehogs, rats and elephants are also susceptible. Human infection is very rare, and results in a mild, self-limiting illness. There are only two recorded cases of FMD in humans, one in Britain in 1966, and another suspected case there in 2001. But FMD is the most important animal pathogen in terms of international trade of animals and animal products, making this disease not a safety issue, but an economic one. For example, an outbreak of FMD in Taiwan in 1997, illustrates how this illness can wreak economic disaster on a country's economy. It necessitated the destruction of eight million pigs, causing Taiwan to lose its primary export market for pork.

FMD is endemic to Asia, parts of Africa, South America and the Middle East, but sporadic outbreaks do occur elsewhere. Through the use of vaccination programs, most developed countries have eradicated FMD, and now ban the importation of certain animal products in an attempt to avoid reintroducing the disease. The FMD vaccine, however, is not allowed by the European Union, and Europe recently experienced the economic impact of a large outbreak of FMD.

In 2001, FMD was found on over 2,000 farms in England, and another 7,500 farms were designated as having susceptible livestock. Possible cases were also reported in France, Ireland and the Netherlands. To halt the spread and hopefully eradicate FMD, it was necessary to destroy three million sheep, over 600,000 cattle, and 100,000 pigs. This outbreak, with costs estimated to be $3.4 to $5.8 billion, affected one quarter of all English farms and economically ruined many farmers. It was traced back to a single farm that had fed garbage to pigs, which is legal in England with certain restrictions: the farmer must have a permit to do so, and the garbage must be ground up and cooked. The garbage suspected of starting this epidemic of FMD, which was fed to the livestock raw, had come from a restaurant that had served illegally-imported meat. FMD is highly infectious and spreads quickly and easily. In this case, it was out of control before English authorities were even aware of a problem. Veterinarians from countries around the world were sent to England to aid in solving the problem, thereby gaining extensive knowledge of the devastating livestock disease and witnessing firsthand the damage it can cause.

Manifestations of Foot-and-Mouth Disease

FMD can spread by various means, but always does so quickly and easily. It is highly contagious and has many variants that enable the virus to stay ahead of the immunity game. Within animal herds it is spread by direct contact, but the causative virus is also capable of traveling 150 miles or more through the air, so nearby farms can easily become infected. Due to current animal transport practices in the U.S., a highly contagious disease such as FMD could spread to as many as twenty-five states in just five days. The FMD virus can also be transmitted by contaminated boots, equipment, or animal by-products. It has an incubation period that ranges from twenty-four hours to ten days.

The early symptoms of FMD are hard to recognize. Dairy cows will decrease their milk production, and all animals will lack an appetite, and develop fevers and increased salivation. Next, ulcers begin to form on the animal's nose, lips and feet. These vesicles will burst open within twenty-four hours, leaving sores that are susceptible to secondary infections, especially on the feet. The illness may be severe enough that the animal will die, but surviving animals may lose hooves and become lame, and females may abort their fetuses. If piglets become infected, they may die suddenly of cardiac failure.

Controlling FMD

According to the FDA, an outbreak of FMD in the U.S. would result in costs far greater than those incurred by England and would cause great losses in food production, processing, and exportation, thereby affecting literally everyone in the country.

It is essential, therefore, to recognize early signs of this disease and eliminate infected animals. This requires primarily education efforts. Veterinarians should be sure farmers are familiar with the early signs of FMD and other exotic diseases, and recommend livestock inspections on a regular basis. If farmers see anything suspicious, they should contact their veterinarians immediately. It is also important that people traveling abroad not bring food or animal products that may harbor FMD or other pathogens back into the U.S. The virus that causes FMD can be destroyed with heat, sunlight and disinfectants. However, there is no known cure for the disease itself.

FMD as a Biological Weapon

Because it is so easily spread, FMD has the potential to be used for terrorist sabotage of America's livestock. If successfully carried out, even a small outbreak could have devastating ramifications if other countries, in response, prohibited import of U.S. livestock or animal products. This was exemplified by the FMD outbreak in Taiwan.

Glanders

The bacterium *Burkholderia mallei*, the causative agent of the disease known as glanders, infects primarily horses, but donkeys, mules, goats, dogs and cats can also become ill with this infection of the upper respiratory tract. The U.S. has not had a case of glanders since the 1940s, but sporadic cases still occur among domestic animals in Asia, the Middle East, Africa, and Central and South America. Humans can occasionally acquire glanders, but it is usually seen among laboratory workers and individuals who are in close, prolonged contact with infected animals. Sporadic cases have been reported among laboratory personnel, individuals who care for horses, and veterinarians. The last recorded human case in the U.S. was in 1945.

Disease Manifestations of Glanders

Animals are the only natural reservoir of this bacterium, and acquire it through inhalation. Humans can also contract glanders this way via contaminated dust or soil, or through direct contact with an infected animal, in which case the bacterium enters through a break in the skin or mucous membranes of the eyes or nose. Person-to-person transmission is also possible, and has been reported among those caring for an infected individual. A few cases have also been acquired through sexual contact.

In humans, glanders may be localized as a cutaneous infection, evidenced by the formation of a pus-filled lesion at the site of bacterial entry, possibly accompanied by swollen lymph nodes. Localized infection could also occur in mucous membranes, and result in excessive mucous formation at the infection site. If the mucous membranes of the nose are infected, nodules will form in the nasal cavity and there may be a bloody discharge. These localized infections are easy to recognize and treat, but if the organism is inhaled, a more serious, pulmonary infection can ensue, wherein abscesses form in the lungs. A case of pulmonary glanders may develop into pneumonia. A third type of glanders, chronic glanders, is a suppurative (pus-forming) infection of the skin that also causes abscesses in muscle tissue of the arms or legs, or in the liver or spleen. Any type of glanders can become a systemic, bloodstream infection. This final manifestation of glanders is generally fatal within a week to ten days.

General symptoms of the various forms of glanders include a fever, muscle aches and tightness, headache, and chest pain. Sometimes patients will have excessive tear formation and light sensitivity, or diarrhea.

Diagnosis and Treatment

B. mallei can be isolated from the blood, urine, skin lesions or sputum of infected individuals. Antibodies can be detected in blood samples, but only after two weeks or more from the time of the initial infection. Currently there is no vaccine for glanders, but luckily the bacterium is sensitive to several types of antibiotics. Infected animal herds must be destroyed. Healthcare workers can avoid infection through the use of standard blood and body fluid protective gear.

Glanders as a Biological Weapon

This agent is of concern because it can infect humans, only a very few organisms are needed to initiate an infection (only one to ten bacteria delivered as an aerosol is lethal) and it has been used as a weapon in the past. During World War I, German operatives spread anthrax and glanders among horses and mules of rival cavalries. *B. mallei* was used again in World War II, this time by the Japanese who infected horses and people in China. The U.S. also studied this organism during World War II for potential use as a biological weapon.

Melioidosis (Whitmore's Disease)

This illness is caused by another bacterium in the genus *Burkholderia*, *B. pseudomallei*. Melioidosis is similar to glanders in its disease manifestations, but this organism differs from *B. mallei* in its ecology and epidemiology. *B. mallei* is found naturally only in animals and in several parts of the world, whereas *B. psuedomallei* is a soil and water contaminant found only in the tropics. It is endemic to and especially prevalent in Southeast Asia, with the highest concentration of cases occurring in Vietnam, Burma, Laos, Thailand, and Malaysia. It has also been found in northern Austrailia, India, the South Pacific, Africa and the Middle East. There have been a few cases in the Western hemisphere as well. The United States reports between zero and five cases each year, but only among travelers to countries where the disease is endemic, or among immigrants from those countries.

Manifestations of Melioidosis

Both humans and animals, including sheep, goats, cats, dogs, horses, cattle and swine, acquire this disease when they come in contact with or ingest contaminated soil or water, or from inhaling dust containing the bacterium. In areas of Southeast Asia, this bacterium is routinely found in agricultural soil where it can also contaminate foods. Humans can also become infected through breaks in the skin, and person-to-person transmission does occur. Two documented cases of melioidosis were spread via sexual contact. Incubation time for this disease can vary from two days to many years.

Disease manifestations of melioidosis are similar to those of glanders. Infections may be localized on the skin, or may affect the respiratory system. The latter type results in a characteristic productive or nonproductive cough. Also similar to glanders, melioidosis may develop into a bloodstream infection that can be fatal, involving respiratory distress, the formation of pus-filled lesions throughout the body, and mental disorientation. Finally, melioidosis can become a chronic infection of the joints, viscera, skin, brain, bones, lymph nodes, liver and spleen.

The general symptoms are basically the same as those seen with glanders and include fever, headache, loss of appetite, diarrhea, and muscle soreness.

Diagnosis and Treatment

B. pseudomallei can be isolated from the blood, sputum, urine, or skin lesions of victims, or antibodies to the bacterium in the patient's blood can be detected and measured. This bacterium is also sensitive to many antibiotics, but treatment must be started early in the infection.

Prevention of melioidosis in areas of the world where it is endemic is difficult because it is so prevalent in water and soil. Persons with underlying medical conditions such as HIV or diabetes are particularly susceptible to infection, and should avoid contact with contaminated sources. People with skin lesions should also exercise caution. Healthcare workers must be protected by wearing appropriate clothing, masks, and gloves, that are typically used to guard against blood and body fluid contamination.

Melioidosis as a Biological Weapon

In and of themselves, neither *B. pseudomallei* nor *B. mallei* would make very good bioweapons because they are readily detectable using simple bacteriological techniques, and are treatable with several common antibiotics. If used as weapons against humans, both organisms would most likely be dispersed as aerosols.

Melioidosis was supposedly used during the Vietnam War to infect about 250,000 U.S. military personnel, and it seems that these organisms are better suited to such use as military weapons rather than as agents to be used against large populations of civilians, if at all.

Crop Pathogens

Biowarfare experts agree that livestock are more vulnerable than crops to agricultural bioterrorism due to the rapidity with which a disease can spread among animals. They also agree, however, that a terrorist attack on the agricultural crops of the U.S. is possible or even probable, but they disagree on how successful such as attack might be. U.S. crops are vulnerable in that they are engineered to grow in a certain climate and under particular environmental conditions, and to be resistant to the types of pathogens ordinarily found in these specific areas. Therefore, say some, these crops may be particularly sensitive to *other* types of pathogens. For instance, North America and Western Europe grow only one or two varieties of major crops; the U.S. emphasizes corn, wheat and soybeans, each of which has been genetically engineered and is therefore quite homogeneous and vulnerable to pathogens not normally found in the U.S.

Others say that most crops are resistant to a wide variety of plant pathogens because crops, unlike livestock, are not grown in isolation. However, even these "resistant" crops may still be susceptible to foreign pathogens, or agents that have been genetically engineered to be particularly virulent or immune to the effects of pesticides.

Opinions also vary regarding the potential effectiveness of disseminating a crop pathogen using crop-dusting techniques, the most likely means of spreading the agent. Some experts say that once it is released, the pathogen would infect the crop, multiply, and continue to spread on its own. Alternatively, others argue that plant disease epidemics are highly dependent on environmental conditions for their spread, and that it is quite unlikely that the technology or methodology for overcoming these environmental effects would be available through the limited means of a small terrorist group.

Professor Larry Madden of the Plant Pathology Department at Ohio State University, in a press release covering the August 10, 2001 symposium in Montreal at the annual American Phytopathological Society and Canadian Phytopathological Society meeting, stated that a crop contaminant, no matter how big or small, could cause a loss of confidence in the safety of the food supply. Professor Madden also pointed out that a single episode could open the door to the proliferation of hoaxes, which could possibly be no less harmful than the real thing. He also offered the following scenario of a terrorist's use of a plant pathogen: It would probably be an agent that has been genetically engineered to be especially virulent to a major crop plant. This agent would be prepared in large quantities and sprayed onto crops using a small airplane, like a crop duster. Millions of spores would be carried on the wind to spread beyond the range of the original distribution, and would be able to survive for a long time in the environment. The initial infections would allow the pathogen to multiply further, producing more spores and proliferating the disease until it reached epidemic proportions. Alternatively, a seed source could be infected—many seeds are grown in other countries, beyond U.S. control, or the soil could be poisoned with root or lower stem diseases. (*Bioterrorism May Be a Threat to U.S. Agriculture, Expert Says, see the website:* www.osu.edu/units/research/archive/croppat1.htm).

Pathogens that would be suitable candidates for crop terrorism would be those that are easily produced, capable of surviving and spreading in the environment, and effective in damaging crops. Even if only small portions of a crop were to be infected, the entire crop may have to be destroyed to prevent the spread of a pathogen. Another concern about a terrorist's use of plant pathogens involves the production of toxins by some types of plants when they become infected with certain pathogens. Even small amounts of these toxins can cause sickness and death in animals and people.

Animal & Plant Pathogens as Biological Weapons

The use of these types of pathogens may be appealing to some terrorist groups for various reasons. The direct taking of human life would be avoided; the sabotage of livestock or crops would be simple with little or no personnel risk to the perpetrator; the public and government reaction to this type of sabotage would be far less intense than an attack upon civilians; deliberate destruction of livestock or crops may also be easier to cover up since it would be hard to distinguish from a naturally-occurring disease outbreak. According to Peter Chalk, a Policy Analyst with the RAND Corporation, in a paper entitled *The Threat Beyond 2000*, cases of agricultural bioterrorism may have been missed in the past for this reason. Chalk's paper also says that most countries do not consider agricultural terrorism to be a punishable criminal act, and there has been little or no interest in including agriculture in counter-terrorism plans.(RAND's, *Bioterrorism: Homeland Defense: The Next Steps* conference: http://www.mipt.org/agterror-rpt.html).

Additional motivating factors may contribute to the choice of agricultural terrorism as a form of aggression. This type of terrorism has the potential to destabilize the economy, to undermine the American people's confidence in government and social stability, and to create mass panic. A terrorist or other perpetrator may use agricultural terrorism, then take advantage of a ruined market for their own economic gain, or use the threat as a means of blackmail.

While these factors make agricultural terrorism seem appealing, it has hardly ever been used. Chalk believes that terrorists may not yet have realized the potential effects of an attack on America's crops or livestock, or may possibly feel that this form of terrorism does not possess the attention-getting capacity of other forms.

Research has shown that the development of effective weapons for use against livestock and crops is not a project to be undertaken by novices. The weaponization of these pathogens requires a dedicated infrastructure, highly trained personnel and significant resources. The pathogen must be acquired, propagated, and processed in a form suitable for delivery. The delivery system itself has to be designed and constructed, and have the ability to overcome environmental conditions, especially in the dissemination of a crop pathogen. In addition, pathogens may require a protective coating to increase survival time in the environment. It would, however, be easy to get foreign animal or crop pathogens by various means. They could be obtained from international laboratories or repositories, isolated from diseased animals or plants, or small quantities could even be carried across unregulated borders or sent in the mail. In most cases, it would be difficult to cause a widespread disease outbreak without using a particularly virulent strain.

A successful attack on agriculture would shake the confidence of the American people in their government's ability to protect them against terrorist acts of any kind. In addition, the destruction of a meat source or major crop would affect not only farmers, food processors and distributors, but every citizen who would feel the economic impacts of greatly reduced food supplies, and the inability of the United States to export food to other countries. Ramifications of zoonotic diseases (those that cane be transmitted from animals to humans) are much greater yet than these.

In addition to revenue losses for the United States, a widespread act of agricultural bioterrorism would mark financial ruin for many farmers who may not be able to bear replacement costs of lost herds and crops, or the cost of decontamination, if necessary. History indicates that other, unrelated industries would be affected as well; due to the foot-and-mouth outbreak in England, British tourism lost about $5 billion.

What Can Be Done to Prevent Agri-terrorism?

In their article *"Planting Fear: How real is the threat of agricultural bioterrorism?"* the authors; Gavin Cameron, Jason Pate and Kathleen M. Vogel state:

> It is extremely unlikely that any agricultural bioterrorist could fatally wound the entire U.S. agricultural sector or national economy, both of which are strong and diversified. Local, regional and national effects, however, could be significant. Even to the extent that the United States is vulnerable, it is unlikely that terrorists could strike successfully. An advanced, state-level bioweapons program, however, might be able to overcome or circumvent some technical hurdles. Given the bioweapons programs of Iraq and the Soviet Union, it is possible that state programs might explore agricultural options in the future.
>
> (Source: *Bulletin of the Atomic Scientists*, September/October 2001, Vol. 57, No. 5, pp 38-44)

To prevent this type of terrorism, the authors suggest an increase in communication efforts among scientists involved in the epidemiology of animal, plant and human diseases, and among veterinarians, noting that disease in wildlife is often a harbinger of human disease. Increased educational efforts among farmers and veterinarians, along with and extension of disease surveillance down to individual farms, would help to increase accurate diagnoses and decrease response times. It is necessary to augment laboratory-testing facilities for rapid diagnosis of animal diseases, and to further research foreign animal and plant diseases. Current agricultural practices that make U.S. livestock more vulnerable to attack should be reevaluated. Finally, methods should be developed to distinguish between natural disease outbreaks and those instigated by terrorist acts, and more attention devoted to the protection of American agriculture.

Animal Pathogens: FMD, Glanders, Melioidosis

Disease	Symptoms	Transmission	Prevention / Treatment	Food / Water Sabotage
Foot-and-mouth disease (virus)	In cattle, swine, sheep etc: decreased milk production, anorexia, fever, increased salivation, ulcers on nose, lips & feet	Direct contact, via aerosols and fomites	Destroy infected animals to prevent spread / No cure	No / No
Glanders (bacterium)	In humans: skin lesions, nodules at point of infection; also pulmonary lesions if inhaled	Inhale contaminated dust or soil; through skin via contact with infected animal; person-to-person	Avoid contact with infected animals etc. / Antibiotics	No / No (aerosol delivery possible)
Melioidosis (bacterium)	In humans: similar to glanders	Inhale contaminated dust; ingest contaminated food or water; via break in skin; person-to-person	Avoid contaminated sources / Early intervention with antibiotics	Possible but not likely / Possible but not likely

PART FOUR

Fighting
Back

CHAPTER FOURTEEN

Working Together

The threat of terrorism looms prominently in every American's mind since the catastrophe of September 11, 2001. Physical threats—planes crashing into buildings, bombs exploding—fear-generating though they may be, are not as terrifying as the possibility that biological and chemical weapons are available

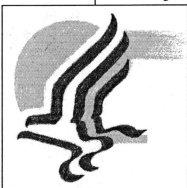

United States Department of Health and Human Services (DHHS) logo

and have been used with some frequency in the past, and may again be used in the future. We can increase airport security and wear gloves and facemasks to handle our mail, but what can we do about the potential for contamination of the very things that sustain us—our food and water?

First and foremost, we must not live in fear. Fear has a crippling effect and could render us incapable of doing anything at all to protect ourselves or our food and water. Secondly, we must put faith in the government agencies that protect these supplies: the United States Environmental Protection Agency, the Food and Drug Administration, and the Department of Agriculture. Finally, the American people must arm themselves with knowledge about biological and chemical agents, how to protect against them and how to recognize symptoms of illness and signs of an intentional release. We must be informed, prepared and proactive.

Food Industry Efforts

The National Food Processors Association (NFPA), in response to the September 11 attack, formed the **Food Security Alliance,** which functions as a forum for information exchange and coordination of preparedness activities. Participants include agricultural and food industry trade associations from production through retail, government agencies responsible for food and water safety, public health agencies, and law enforcement. A *Food Security Checklist* for use by suppliers and processors is available through the NFPA to assess security in individual operations.

The Federal Regulatory System of Food Safety

The current United States regulatory system for ensuring food quality and safety, made up of agencies that function at the local, state and federal levels, has four primary functions:
- establishing safety standards and tolerance levels;
- monitoring and inspecting of farms, storage, processing and retail establishments;
- tracking and intercepting commodities or shipments determined to be "high risk";
- stopping or quarantining international and domestic shipments, rejecting food lots, shutting down plants, assessing penalties and prosecuting suspected offenders; and
- tracking food safety problems by tracing, recording and analyzing reports of illnesses, outbreaks, and deaths attributed to food safety problems.

The Food and Drug Administration (FDA)

The Food and Drug Administration within the Department of Health and Human Services ensures the safety and wholesomeness of all food sold in interstate commerce except meat, poultry and certain egg products, which are under jurisdiction of the United States Department of Agriculture. The FDA operates at all levels of government—federal, state and local—monitoring the manufacture, importation, transport, storage and sale of these foods. The FDA also develops standards for food composition, quality, nutritive value, safety, and labeling. It conducts its own research on methodologies to detect and prevent contamination, and collects and interprets other data on nutrition, additives, and pesticide residues. Probably the most well-known activity of the FDA is various aspects of food inspection, including processing plants and imported products, but the FDA also regulates radiation-emitting products such as microwave ovens, and enforces pesticide tolerance levels established by the Environmental Protection Agency. It monitors seafood along with the National Marine Fisheries Service. The FDA's Center for Veterinary Medicine monitors animal feeds and safety of food from animals, with focus on feed for livestock and poultry destined for human food. Such monitoring can prevent spread of bovine spongiform encephalopathy (BSE) and other diseases.

In November of 2001, the Secretary of Health and Human Services stated that the FDA is devoting its full attention to counter-terrorism. Additional inspectors have already been hired, some of whom were deployed to ports. Other efforts will include enhancement of international inspections and development of new methods to detect contaminated foods. Since September 11, 2001, the FDA has increased its emergency response capabilities, and continues to work with other federal, state and local food safety authorities and with regulatory agencies abroad to protect and to respond rapidly to evidence of threats. On June 12, President Bush signed *The Public Health and Bioterrorism Preparedness and Response Act of 2002,* which designates billions of dollars for terrorism protective measures, some of which will go to the production and stockpiling of smallpox vaccines and hospital preparedness. The bill is divided into five sections, each dealing with a different aspect of public health and safety, including the control of dangerous biological agents and toxins, and the protection of our food and water supplies.

Protection of the food supply falls under FDA jurisdiction and deals with various aspects of food safety. Included are strategies for crisis communication, as well as the education of food industry personnel so they are able to assess possible threats in the manufacturing or transportation of foods and to implement protective measures. Adulteration of foods is also covered, with an emphasis on rapid tests and sampling procedures for imported foods. Facilities that manufacture, process, pack or hold food must be registered with the FDA. The FDA must also be allowed access to pertinent records when there is reasonable belief that an article of food is adulterated and presents a threat of serious adverse health consequences or death to animals or humans. Importers must give notice prior to the arrival of food at ports of entry, and the FDA has authority to refuse entry of questionable foodstuffs.

The FDA and the food industry's Food Security Alliance have worked together toward the goal of increasing the physical security of food plants. To that end, they developed, and FDA issued in January 2002, *The Food Security Preventative Measures Guidance* to enable manufacturers, processors, packers, distributors, and retailers to better protect our food from terrorism. However, the measures outlined in this document are, at this point, only voluntary. They are supplied to processors, producers, transporters, and retailers of food, and contain recommendations for increasing security. These recommendations include:

- checking criminal backgrounds and immigration status of employees;
- watching out for employees exhibiting odd behavior, such as staying at work unusually late or trying to obtain access to files, etc;
- enhancing plant security measures; and
- developing guidelines for restricting access of certain areas to authorized personnel.

The FDA states that these guidelines are common-sense measures that refocus attention from accidental bacterial contamination scenarios to those involving intentional tampering, but it will take effort and money for food companies to implement changes and increase security measures. (These guidelines are available on FDA's web site, www.fda.gov).

The Centers for Disease Control and Prevention (CDC)
The Centers for Disease Control and Prevention form an agency of the United States Public Health Service devoted to preventing disease, disability and premature death. It consists of various divisions that work with state and local health departments, academic institutions, and various other organizations to keep tract of environmental and chronic health problems as well as food-borne illness. Besides investigating outbreaks, the CDC monitors food-borne illnesses to gain information on early warning signs of impending outbreaks. They keep track of progress made in

reducing the incidence of food-borne illness, and make observations on new or changing patterns of outbreaks. If a food product is implicated in an outbreak, the CDC alerts and works with the FDA and the USDA to protect the public. The CDC also assists state and local departments in developing epidemiology for other types of diseases and for outbreak response strategies.

The CDC has been actively pursuing biological terrorism preparedness since 1998, but the agency's initial actions were not taken until Fall 2001, in response to anthrax-contaminated mail. In February of 2002, the CDC commented that the current state of public health infrastructure in this country is not adequate to detect and respond to a large-scale bioterrorist attack. Therefore, the CDC initiated the development of a *Bioterrorism Preparedness and Response Program* that integrates planning and training, public health preparedness including surveillance, epidemiology, rapid laboratory diagnosis, emergency response, and communications systems. Through the Centers for Disease Control and Prevention, the Department of Health and Human Services has allocated millions of dollars for increasing preparedness at the state and local level, which amounts to an almost ten-fold increase from 2001. (The 2003 level should be the same or higher.)

The CDC has identified the biological agents most likely to be used for terrorism and is working with states to assist in the detection and medical management of exposure to these agents. Health care providers must be educated to recognize symptoms of illnesses that could arise due to use of a biological or chemical agent such as anthrax, smallpox, plague, tularemia, botulism, or viral hemorrhagic fever. To that end, public health workers and lab personnel will be trained to recognize symptoms of these diseases. They will be able to answer questions such as:

- What does the syndrome look like?
- What tests need to be ordered?
- Is isolation necessary?
- How do you report it?

CDC also helps state and selected local public health departments improve their preparedness and response capabilities for bioterrorism.

Surveillance systems already in place for emerging infectious diseases serve as the foundation for terrorism preparedness and response, but surveillance of respiratory distress syndrome and hemorrhagic fever or meningitis symptoms will be increased in order to recognize possible intentional use of a disease agent. This surveillance includes emergency room visits, monitoring of pharmacy records for use of antidiarrheals or antibiotics, monitoring 911 calls and calls to poison control centers, and strengthening links to the veterinary community.

The CDC maintains that prompt identification of the agent is the most important step when a threat arises, whether due to natural causes or a deliberate act of terror. To strengthen their ability to do this, the CDC has taken steps to enable regional and state laboratories to handle more samples and to better communicate findings. It points out that public health must begin at the state and local level, and expand to the regional and national level. Once an agent is identified, victims can be properly treated and exposure to others can be limited. The CDC has established a *Rapid Response and Advanced Technology Laboratory* that can provide rapid identification of naturally-occurring biological agents that are rarely seen in the United States.

While laboratory expansion efforts will continue into the future, the number of personnel employed by the CDC has been increased as well, to include health communication experts. Establishment of an *Emergency Operations Center* has increased CDC's ability to work in a timely manner with affiliates around the nation and the world. CDC's chemistry laboratory enables scientists to evaluate 150 toxic chemicals in human beings that could be used as chemical weapons. The agency has also worked with pharmaceutical companies to stockpile antibiotics, antitoxins and vaccines, to ensure that they can be made readily available in time of crisis. Stockpiles can be relocated to areas of the U.S. in need within twelve hours.

Future steps that the CDC plan to take include:
- continued enhancement of public health infrastructure for bioterrorism response;
- continued development of response capacity;
- training in bioterrorism preparedness and response for the public health workforce; and
- continued enhancement of the national pharmaceutical stockpile and information systems.

The United States Department of Agriculture (USDA)

The United States Department of Agriculture's *Food Safety and Inspection Service* monitors domestic and imported meat, poultry, and certain egg products for bacteria, drugs, chemicals, and pesticides. The *Agricultural Marketing Service* performs food quality functions such as commodity standardization, inspection and grading. The *Federal Grain Inspection Service* inspects corn, sorghum and rice for aflatoxin and checks the quality of domestic and exported grain, rice and related commodities. In the past, the USDA has worked to prevent the unintentional introduction of pests and diseases. These efforts have now expanded.

On October 8, 2001, President Bush established the *Office of Homeland Security.* The USDA then established the *Homeland Security Council* to work in partnership with Homeland Security, the National Security Council and other departments. The Council will establish policy, coordinate security issues, track USDA's progress on objectives, and ensure that information is shared and activities are coordinated with other federal agencies. There are three sub-councils of the Homeland Security Council that deal with protecting USDA facilities and infrastructure, protecting USDA staff and Emergency Preparedness, and protecting the food supply and agriculture production. This last group addresses issues dealing with food production, processing, storage and distribution. It evaluates threats against the agriculture sector and develops rapid response measures to such threats. Additionally, it is involved in border surveillance and protection to prevent introduction of plant and animal pests and diseases, and in food safety activities concerning meat, poultry, and egg inspection.

As part of the Public Health Security and Bioterrorism Preparedness and Response Act of 2002, the Department of Health and Human Services will be working with the USDA to regulate possession and use of *Select Agents*, defined as hazardous organisms and toxins used in research. Examples of these agents are *Bacillus anthracis, Yersinia pestis*, and the Ebola virus.

The Administrator of the Food Safety and Inspection Service (FSIS) of the USDA, and the Director of the Center for Food Safety and Applied Nutrition of the FDA, co-chair *PrepNet*, a group that focuses on rapid response and preventative activities to protect the food supply. PrepNet has three subgroups that will focus on emergency response, laboratory capability, and efforts aimed at prevention and deterrence.

After September 11, 2001, the USDA's Under-Secretary for Food Safety formed the *Food Biosecurity Action Team* (F-BAT). F-BAT will coordinate activities of biosecurity, counter-terrorism, and emergency preparedness within the Food Safety and Inspection Service (FSIS), and serve as the voice on biosecurity with other agencies and constituents. FSIS will work closely with the CDC, FDA, and EPA as well as state and local agencies, sharing information about illnesses to prevent biosecurity threats.

In January of 2002, the United States government allocated almost $300 million for:

- improving the physical and operational security at key USDA locations and the USDA's Plum Island Laboratory;
- the Agricultural Quarantine Inspection program, whose efforts restrict the entry of pests and diseases at the borders;
- for the Food Safety and Inspection Service to enhance monitoring, training of inspectors, hiring of additional inspectors, and to expand technical capabilities; and
- improving rapid detection methods for animal diseases through the Agricultural Research Service.

The USDA also allocated grant and federal/state partnership monies to:

- help increase preparedness in individual states;
- improve surveillance for animal diseases;
- improve detection of plant and animal diseases;
- rapid and accurate diagnosis of animal diseases; and
- strengthen response capabilities.

The Environmental Protection Agency (EPA)

The *Environmental Protection Agency* was established in 1970, and in May of 1998, was designated to oversee protection of public water supplies. The EPA works with state government to establish permissible levels of contamination for seafood-harvesting waters. It regulates new pesticides by reviewing all scientific data, registering the pesticide only if it will not pose significant risks to human or animal health, or to the environment. It also regulates the distribution, promotion, handling, storage, use, and disposal of pesticides, and sets tolerance limits that are enforced by the FDA and the USDA. The EPA can also revoke the use of pesticides if necessary, based on new findings.

Threats to our water supplies include chemical, radiological or biological agents, damage or sabotage of the physical infrastructure, or disruption of computer systems. According to the EPA, the chances of a terrorist contaminating drinking water is reduced by several factors, including the diluting effect encountered when introducing an agent into a large quantity of water, and treatment systems currently used (chlorine, filtration) that will deactivate many contaminants. The EPA contends, however, that work needs to be done to assure that all public water supplies are adequately protected.

Following September 11, the nation's water utilities were instructed to look at their security systems and augment surveillance and protective measures, and in October of 2001, the *Water Protection Task Force* was established to protect and secure water supplies. The EPA worked with the FBI to contact local law offices, asking them to work closely with their local water utilities to provide extra security. Together with its partners in the private sector, the EPA developed training for utilities on how to assess vulnerability and determine actions to guard against attack, and develop emergency response plans. The EPA continues to provide information regarding steps water facilities can take to protect their sources of supply and their infrastructure.

The National Marine Fisheries Service
The National Marine Fisheries Service of the Department of Commerce is responsible for the quality and identification of seafood, fisheries management and development, habitat conservation, and aquaculture. This service inspects and certifies seafood-processing plants, and is responsible for ensuring the safety of seafood with help from the FDA and various state agencies.

State-Level Regulation
At the state level, regulatory agencies include boards of health, departments of human and social services, state universities, and environmental and sanitation agencies. States enforce laws and standards that coincide with federal laws, and work with federal

agencies to safeguard the food supply. For example, the FDA field staff works with state authorities to obtain and evaluate samples, and to exchange information. Some of the major agricultural states have their own inspection systems with strict safety standards, and some states monitor the safety of their produce.

At the Local Level

All aspects of food production—from farmers and producers, processors and manufacturers, to shippers, packers and retailers— share in the responsibility of safeguarding the food supply. Every sector of the food industry must adhere to government regulations regarding the use of pesticides, animal husbandry practices, administration of animal drugs, food plant sanitation, food processing procedures, and institution of quality control programs such as Hazard Analysis and Critical Control Points (HACCP). The research and development (R & D) departments of food processors and manufacturers develop and improve products, and assure the safety of all food ingredients. Food manufacturers and processors have their own quality control and quality assurance programs, and end products are often tested by outside laboratories as well. Farmers work with government extension agents and land-grant universities to stay informed on production and safety practices for farms. Veterinarians and consultants help to prevent pesticide damage, and advise on animal health care. Finally, the government and the food industry are involved in various educational efforts for people employed by the food industry, as well as for the general public.

Preparing For and Recognizing Terrorism
Directed at Our Food or Water

The Centers for Disease Control and Prevention contends that if our food or water supplies were to become targets for terrorists, the agents of choice would most likely be those with proven track records, such as certain chemicals or *Salmonella* bacteria, which have all been used in the past to purposely contaminate food or water. The CDC also stresses that complacency must be avoided—the United States needs to be prepared with emergency response plans that have been carefully formulated and evaluated. Even though the U.S. has excellent surveillance systems already in place, these systems should be strengthened in light of the new threat of terrorism. The capacity of epidemiological laboratories also needs to be enhanced to rapidly detect food-borne illness and link an outbreak to its source.

According to the CDC, early detection of sabotage is essential. Clinicians must be able to recognize and respond to suspicious or unusual symptoms, and to communicate findings to determine an increase of these presentations that may be above the norm. A bioterrorist attack would probably differ from a natural disease outbreak in that it would be a non-endemic agent; the outbreak could occur without warning at any time of year, as opposed to naturally-occurring outbreaks which are seasonal; and the number of victims could be large, depending on the agent and mode of transmission.

To better prepare for such an attack, the CDC recommends:

- maintenance of effective disease surveillance through a nationwide increase in epidemiologic and laboratory capacity;
- plans for rapid identification and characterization of agents;
- emergency distribution of large quantities of medical supplies, especially antibiotics and vaccines;
- strengthening and coordination of communication to minimize response time, especially early on when exposed yet asymptomatic persons may still be treated prophylactically;
- shorter response time to increase the possibility of apprehending perpetrators; and
- education and training in bioterrorism and its potential consequences.

Vaccination Programs

The CDC indicates that the high cost and inherent difficulties of vaccinating large populations of U.S. citizens, along with the broad spectrum of potential terrorist agents, makes a nationwide vaccination program an unlikely means of protection. While vaccines, therefore, cannot be used in this way, they *can* be used to control an epidemic and subsequently prevent a global pandemic. They can also be used for post-exposure prophylaxis when necessary, as with anthrax in the mail, or pre-exposure prophylaxis for high-risk groups such as first responders, clinicians and laboratory personnel.

The U.S. government is working on increasing vaccination programs, developing and stockpiling smallpox and other vaccines, as well as assuring an adequate supply of antibiotics. It is also studying sensors for early detection of a biological attack, and looking for ways to clean up or neutralize released agents. Emergency response teams will be prepared to respond to biological agents, and doctors trained to deal with them.

What You Can Do

Drinking Water

The EPA contends that it is unlikely that a small amount of a biological or chemical agent could effectively contaminate a city's water supply. Extremely large amounts of a contaminant would be necessary, and it would be difficult for a saboteur to accomplish this without being noticed, especially with the increase in security at reservoirs and utilities across the country. And as previously stated, chlorination and other treatments would remove or inactivate many agents.

The EPA has the following recommendations regarding drinking water: bottled water may not be safer that tap water since it may come from a source that is just as vulnerable to attack as your local water utility. The safety of bottled water also depends on security measures at the bottling plant. Tap water is protected at the local level through security measures advocated and supported by the EPA, state and local government, and state- and locally-based water organizations. Boiling of drinking water will help destroy microorganisms, but could also concentrate certain contaminants such as lead.

If your local drinking water supply becomes contaminated, the utility will activate its emergency response plan in conjunction with local law enforcement and state emergency officials, which provides for shutting down the system, notifying the public of emergency steps that should be taken, and providing and alternative water source, if necessary.

If something or someone looks suspicious around your drinking water supply, contact the police or call 911. The Safe Drinking Water Hotline is 1-800-426-4791. If you believe you have witnessed terrorist activity, call authorities and be prepared to provide detailed information.

Foods

Consumers should always check their groceries for evidence of tampering, and look for unusual odors, appearance or taste. If the package is damaged or has been opened, do not use it. Call your FDA district office, or the USDA Meat and Poultry Hotline (1-800-535-4555) if you encounter anything suspicious.

As a general rule of safety, observe safe food handling practices in your home. Fresh fruits and vegetables should be washed with plenty of clean water. Utensils, cutting boards and hand should be washed with hot, soapy water. Raw meats should always be kept separate from ready-to-eat foods to prevent cross contamination. Foods should be cooked thoroughly, leftovers cooled promptly and reheated completely.

Do not use antibiotics prophylactically unless prescribed by your physician. Keep in mind that antibiotics are not effective against viruses, chemicals, or radiation, only some bacteria.

General Precautions

Experts on bioterrorism agree that agents most likely to be used in an attack would be those that are most dangerous when released as aerosols. If you believe you are in a contaminated area, leave that area immediately. If it is a room, shut the door and turn off ventilation units. If possible, shut down the air handling system for the entire building. Call for emergency help. If you come into contact with a chemical or gas, cover your mouth and nose and leave the area. Thoroughly rinse your eyes and exposed skin with water, and remove contaminated clothing.

Report any suspicious activities you witness, or any suspicious substance you may run across, to your local authorities. Do not try to clean up the substance, but rather cover it, leave the area and close the door, wash your hands with soap and water, and remove contaminated clothing. Report suspicious mail. Without opening the package or envelope, put it in a plastic bag etc. or cover it up and leave the area.

Becoming Proactive

You have already taken the first step in the battle to protect our food and water against terrorism: by reading this book, you have become knowledgeable about the potential agents of food and water sabotage. You have learned about the symptoms of the various infections and intoxications, and learned ways to protect yourself and your family.

This is a good start, but it cannot be the end of your efforts. As American citizens, it is our duty to make sure that those government agencies that are responsible for ensuring the safety of our food and water have the resources to not only maintain their current levels of surveillance, but to *increase* them in light of this new and very real threat.

Your actions can begin at the local level. Contact your government representatives and voice your concerns, stressing that the Food and Drug Administration, the United States Department of Agriculture, the Environmental Protection Agency and other agencies cannot adequately protect our food and water from terrorism without the funds to take actions necessary to do so.

Contact food manufacturers in your neighborhood and ask them if they have increased security measures at their facilities. Do the same at your local water municipality. Have steps been taken at your own clinics and hospitals that will enable them to respond adequately to a terrorist attack in your area?

Finally, seek advice and answers from the authorities to further expand your knowledge base. Terrorism is here to stay, but what form it will take has yet to be seen. We have taken for granted the safety of the food we eat and the water we drink, but we can do so no longer, for they are vulnerable. *We are vulnerable,* but nevertheless, have many options to protect ourselves and our families.

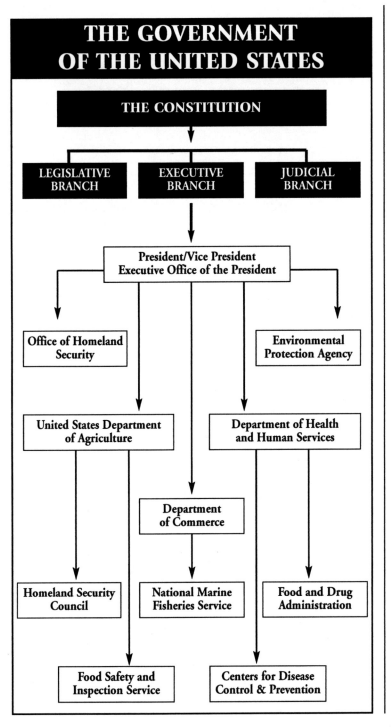

THE GOVERNMENT OF THE UNITED STATES

THE CONSTITUTION

LEGISLATIVE BRANCH

EXECUTIVE BRANCH

JUDICIAL BRANCH

President/Vice President
Executive Office of the President

Office of Homeland Security

Environmental Protection Agency

United States Department of Agriculture

Department of Health and Human Services

Department of Commerce

Homeland Security Council

National Marine Fisheries Service

Food and Drug Administration

Food Safety and Inspection Service

Centers for Disease Control & Prevention

GLOSSARY

Aerobic: requiring oxygen, i.e., aerobic bacteria.

Aerosol: airborne; substance delivered as a spray.

Aflatoxin: toxin produced by *Aspergillus* mold species that most commonly grow in moist nuts and grains, especially corn and peanuts. They have been proven to cause cancer in mammals.

AIDS (auto immune deficiency syndrome): a disease caused by the HIV virus that results in defective functioning of the immune system with an increased susceptibility to infections and cancer.

Alimentary toxic aleukia: shrinkage of bone marrow due to a disturbance in blood production and maintenance caused by ingestion of toxin produced by the grain mold, *Fusarium*. This disease can be fatal.

Anaerobic: able to live without oxygen; strict anaerobe: cannot live in the presence of oxygen, i.e. *Clostridium* and *Bacillus* bacterial species.

Aneurism: an abnormal pouching and weakening of a blood vessel wall.

Anthrax: a fatal disease of cattle and sheep caused by the bacterium *Bacillus anthracis*. Humans contract by inhaling, ingesting or through skin abrasions.

Antibodies: proteins produced by the immune system in response to a viral or bacterial infection, or due to immunization. Antibodies are very specific.

Antibiotic: drug therapy given to treat bacterial infections. Natural antibiotics (streptomycin, penicillin) are produced by certain bacteria to inhibit other bacteria.

Antibiotic resistance: a bacterium's ability to overcome the effects of an antibiotic. This ability can be acquired through plasmid transfer.

Antigen: a specific marker on the surface of a microorganism that is recognized by the immune system as foreign. Antigens are also used in classification and identification schemes.

Antitoxin: a drug given to counteract a specific toxin, i.e., botulinum toxin.

Antiviral: a drug used to inhibit viral replication to lower the number of virus particles.

Arthritis: a chronic inflammation of the joints with stiffness, pain, and loss of flexibility/mobility, possibly deformity and weakness.

Arthritis, reactive: arthritis that develops as a result of the presence of bacterial antigens.

ATCC (American Type Culture Collection): organization located in Virginia from which bacterial cultures may be obtained for use in research etc.

Bacillus: a rod shaped bacterium; the bacterial genus *Bacillus*, which contains *Bacillus anthracis*.

Bacteria: a one-celled microorganism lacking a defined nucleus.

Bacteremia: bacteria in the blood.

Bacteriophage (phage): virus that infects bacteria.

Black death: a variety of bubonic plague characterized by dark patches on the skin.

Botulism: a lethal food poisoning caused by ingestion of a neurotoxin produced by the bacterium, *Clostridum botulinum*. Paralysis of respiratory muscles leads to suffocation and death.

BSE (Bovine spongiform encephalopathy): degenerative brain disease of cattle that results in sponge-like holes in the brain. Similar to Creutzfeldt-Jakob disease in humans.

Bubonic plague: caused by the bacterium *Yersinia pestis,* transmitted by the bite of an infected rat flea. So-called due to the formation of buboes, particularly in the groin area.

Buboe: area of painful swelling, particularly in the groin, due to infection with *Yersinia pestis* (bubonic plague).

Campylobacter jejuni: common food-borne bacterial pathogen that causes diarrhea.

Capsule: protective outer layer formed by some bacteria; antigenic.

Carrier: someone who harbors an infectious agent but does not experience illness.

CDC (Centers for Disease Control and Prevention): government agency in Atlanta, Georgia, that studies disease patterns in order to better protect the public.

Cellular lysis: the breaking open of a cell; results in death of the cell.

Chromosome: composed of two strands of DNA; located within the nucleus of a human cell, 46 chromosomes carry genetic information (genes) for all characteristics of an individual. Bacteria have only one free chromosome (no nucleus).

Chronic disease: a continuously present illness such as high blood pressure.

Ciguatera: toxin produced by reef fish that feed on certain poisonous dinoflagellates.

CJD (Creutzfeldt-Jakob disease): a degenerative brain disease of humans, related to BSE in cattle, where sponge-like holes form in brain tissue. Variant CJD (vCJD) is specifically associated with the eating of beef infected with BSE.

Clostridium: genus of rod-shaped, strictly anaerobic bacteria that contains the pathogens *C. botulinum* and *C. perfringens.*

Cocci: a bacterium that is spherical in shape, i.e., staphylococci.

Contagious: refers to a disease that can be passed from one person to another.

Contaminate: the accidental or purposeful introduction of a foreign microbe, chemical, etc. to food, water, a surface or other entity where it does not naturally occur.

Crohn's disease: chronic inflammation of the intestine.

Cross-contamination: the spreading of a contaminant from one area to another.

Cryptosporidium: protozoan parasite that is generally water-borne; pathogenic to humans, causes diarrhea.

Culture: in bacteriology, the process of growing bacteria to large numbers, usually for the purpose of study or identification; the resulting quantity of bacteria.

Cutaneous: of the skin.

Dehydration: excessive water loss from the body, usually as a result of diarrhea and/or vomiting. It is the cause of death due to many viral and bacterial illnesses if it is not treated with replacement of fluids and electrolytes.

Diarrhea: urgent and frequent evacuation of watery fecal material, often associated with a gastrointestinal illness.

Dinoflagellates: marine protozoans having two flagella; they are the chief constituents of plankton, and some are responsible for red tides.

Disease outbreak: several people suffering the same symptoms in a given area at a given time, usually due to common exposure to a certain pathogen.

DNA (Deoxyribonucelic acid): the chemical from which genes are made.

Dyspnea: difficult breathing; shortness of breath.

Dysentery: profuse, bloody diarrhea that also contains mucous.

E. coli: a small, rod shaped bacterium that is a normal inhabitant of the intestinal tract, though some types are pathogenic. Serotype O157:H7 causes hemorrhagic colitis.

ELISA (enzyme linked immunosorbent assay): detection assay based on reactions between antigens and antibodies.

Emerging disease: a disease that is currently increasing in incidence within a given region, for example, Ebola in Africa and West Nile in the United States.

Encephalitis: inflammation of brain tissue typically caused by a virus (i.e., West Nile), but can also result from bacterial infections.

Endemic: in context, refers to a pathogen or disease that is common to a particular area, i.e., Ebola virus is endemic to Africa.

Endocarditis: inflammation of the tissues lining the heart.

Endospore (spore): protective, inert form of bacteria, characteristic of members of the genera *Bacillus* and *Clostridium*, that allows these organisms to survive adverse conditions.

Endotoxin: a toxin that is not released into the environment, but is a component of the cell's surface or interior.

Enteric: relating to the intestinal tract.

Enterotoxin: a toxin that affects the intestinal tract.

EPA (Environmental Protection Agency): an independent agency that oversees water quality, pesiticides and other issues.

Epidemiology: the study of disease epidemics.

Equine: relating to horses; the horse family.

Ergotism: a disease caused by eating the toxin ergot, which is produced by *Claviceps* mold species that infect grains and grasses; also called Saint Anthony's Fire.

Exotoxin: a toxin that is released from the producing cell into the environment.

Febrile: referring to an illness that has fever as one of the symptoms.

Fecal material: excrement (feces) from the bowel.

FDA (Food and Drug Administration): government agency responsible for monitoring the safety of non-meat food items and drugs.

Flaccid paralysis: the type of paralysis caused by botulinum poisoning wherein muscles remain relaxed due to inhibition of nerve impulses.

Flagella: whip-like appendages possessed by some bacteria; used for locomotion.

Food-borne: associated with food; a food-borne illness is due to eating foods that are contaminated with a chemical or pathogen.

Fomite: an inanimate object that acts to spread disease.

FSIS (Food Safety and Inspection Service): division of the United States Department of Agriculture that deals with food safety and inspection.

Fusarium mold: toxic mold species that infects grain; when toxic grain is ingested, alimentary toxic aleukia may result.

Gas gangrene: disease caused by *Clostridium perfringens*, which grows anerobically, producing gas that destroys tissue.

Gastroenteritis: a disease of the gastrointestinal tract, usually with symptoms of nausea, vomiting, and/or diarrhea or dysentery.

Gastrointestinal: referring to the various regions of the intestinal tract which include the esophagus, stomach, small intestine and large intestine.

Giardia lamblia: protozoan pathogen responsible for giardiasis, an illness manifested by diarrhea, cramps and weight loss; usually water-borne.

Glomerulonephritis: inflammation of renal capillaries.

Guillian-Barré syndrome: a rare disorder that may result from an infection wherein a person's immune system attacks the body's own nerves.

HACCP (Hazard Analysis Critical Control Point): preventative system used by food plants to identify potential microbiological, physical and chemical hazards and implement measures to protect against them.

Hanta virus: transmitted by rodents; causes severe hemorrhagic symptoms including internal bleeding, kidney failure and death.

Hemolysis: destruction of red blood cells.

Hemorrhage: bleeding.

Hepatitis: a viral disease that causes inflammation of the liver.

Histamine: in context, an end product of bacterial breakdown of the amino acid histidine; when ingested, histamine causes tissue swelling, hives and itching.

HUS (Hemolytic Uremic Syndrome): a disease resulting in anemia wherein red blood cells are destroyed at a rate faster than that which they can be replaced; a potentially fatal manifestation of infection with E. coli O157:H7 that typically affects young children and often requires kidney dialysis and blood transfusions.

Hypotension: low blood pressure.

Immune system: complex system that combats infections, toxins and allergies.

Immunocompromised: condition wherein the immune system is not functioning properly, due to i.e., organ transplant, pregnancy, cancer, AIDS, making the individual more susceptible to disease.

Immunohistochemical staining: diagnostic technique used to detect pathogens.

Immunization (vaccination): the introduction of a nonpathogenic form of a virus or bacterium, causing the production of antibodies that will protect the individual from future infections with that pathogen.

Incubation time: the time that elapses between exposure to a disease agent and onset of illness.

Infection: illness caused by the presence of the microorganism itself as it multiplies within the body (as opposed to an intoxication).

Infective dose: the number of bacterial or viral particles necessary to initiate illness.

Jaundice: yellowing of the skin and whites of the eyes due to bile accumulation resulting from liver disease.

Kuru: form of human spongiform encephalopathy among tribes of Papua New Guinea, who contracted the disease from eating brain tissue of diseased tribal members.

Leptosporiosis: infections with *Leptospira* bacteria.

Lymph: colorless fluid containing white blood cells; bathes tissues.

Lymph node: oval-shaped structures that are part of the lymphatic system; contain mature lymphocytes that fight bacteria, viruses and other foreign invaders.

Macrophage: "large eater"; a phagocytic cell of the immune system.

Mad cow disease: see BSE.

Mastitis: in context, inflammation of the udder or breast typically due to bacterial infection.

Meninges: tissue surrounding the brain and spinal column.

Meningitis: inflammation of the meninges caused by a viral or bacterial infection.

Mollusk: marine invertebrate possessing a shell, i.e., shellfish or snail.

Mucosal tissue/mucous: specialized cells lining body cavities that are open to the environment, such as the nose, gastrointestinal tract, respiratory tract and urogenital tract; these tissues produce a sticky mucous that traps microorganisms to prevent infection.

Mutation: a change in genetic information that either occurs naturally or can be induced, i.e., with radiation.

Mycotoxin: toxin produced by certain mold species.

Nasal: relating to the nose.

Nasopharyngeal: the upper area of the throat (pharynx), closely associated with the back of the nose.

Nausea: queasiness of the stomach with the associated urge to vomit.

Necrosis: tissue death.

Neurotoxic shellfish poisoning: due to consumption of shellfish from Atlantic waters that have eaten poisonous dinoflagellates; when these shellfish are ingested, an illness affecting the central nervous system results.

Neurotoxin: a toxin that affects the central nervous system.

Night soil: human fecal material used as fertilizer.

Norwalk virus: common food-borne virus that causes diarrhea; also called small round structure virus (SRSV).

Oocyst: thick-walled, environmentally resistant cyst; the infective form of *Cryptosporidium parvum* that is ingested to cause illness.

Oropharyngeal: relating to the area at the back of the throat.

Pandemic: disease affecting a large proportion of the population in a given area.

Paralytic shellfish poisoning: syndrome contracted from eating shellfish that have fed on toxic dinoflagellates associated with "red tides;" neurotoxin causes progressive paralysis.

Parasite: organism living in or on another organism that does not provide any benefit to its host.

Pasteurization: heat treatment designed to destroy pathogenic organisms in milk, juice etc.

Pathogen: organism that causes disease.

Pathogenic: capable of causing disease.

PCR (Polymerase chain reaction): diagnostic technique wherein small fragment of DNA are magnified so as to be easily characterized.

Phagocytosis: engulfing of foreign invaders (viruses, bacteria) by specialized cells of the immune system.

Pharynx: area between the mouth cavity and the esophagus; back of the throat.

Plague: epidemic causing high mortality; disease caused by *Yersinia pestis*.

Plasmid: circular piece of DNA that is not part of the chromosome; can be transferred between bacterial cells; often encodes for virulence traits.

Pneumonia: inflammation of the lungs often caused by viral or bacterial infection.

Prion: newly discovered disease agent (protein) that is thought to cause transmissible spongiform encephalopathies such as BSE and vCJD.

Prophylaxis: preventative measures, i.e., vaccines or antibiotics taken to prevent disease.

Protozoa: single-celled organism that has a nucleus, i.e., *Giardia* and *Cryptosporidium.*

Reiter's syndrome: syndrome manifested with arthritis, eye irritation, painful urination.

Ricin: toxic component of castor beans.

Rotavirus: spherical virus; common cause of diarrhea, especially in children.

RNA (ribonucleic acid): cell component involved in protein synthesis.

Saint Anthony's Fire: see *ergotism.*

Salmonella: genera of rod-shaped, pathogenic bacteria that cause typhoid fever and gastroenteritis (salmonellosis); common contaminants of certain foods, especially poultry and eggs.

Saprophyte: organism that obtains nutrients from the breakdown of organic matter.

Scrapie: transmissible spongiform encephalopathy of sheep.

Scrombroid poisoning: due to consumption of fish that have been subjected to bacterial action resulting in production of histamine.

SEB (Staphylococcal enterotoxin B): one of many toxins produced by *S. aureus.*

Sepsis (septicemia): systemic illness caused by a disease agent or its products that have spread throughout the body to create a toxic condition.

Serotype: group of related organisms with common antigens.

Serum: the liquid component of blood.

Shigella: genus of rod-shaped pathogenic bacteria that cause dysentery.

Staphylococcus aureus: pathogenic bacterial species that causes food poisoning (intoxication), skin infections (pimples) and other illnesses; potent toxins produced by this organism cause toxic shock.

Sterile: in context, free of microorganisms.

Strain: organisms of the same species that are further classified based on genetic similarities.

Superantigen: toxin that can trigger a massive immune or autoimmune response.

Systemic: affecting the entire body.

Tick: bloodsucking arthropod that often vectors disease.

Toxin: poison.

Transmission: in context, the means by which an infectious agent is spread.

Traveler's diarrhea: diarrheal illness associated with travel to foreign countries; usually due to consumption of food or water contaminated with certain *E. coli* or other pathogens.

TSST (toxic shock syndrome toxin): one of many toxins produced by *S. aureus*.

Unpasteurized: not subjected to pasteurization.

Vaccine/vaccination: a harmless form of a pathogen used to stimulate an immune response (formation of antibodies) that will protect against future infection with that organism.

Vegetative cell: the form of bacteria that divides (replicates) and carries out other processes associated with life; compare to endospore.

***Vibrio*:** small, curved rod-shaped bacterium, species of which cause cholera and other diarrheal illnesses.

Virulence factor: characteristics of a pathogen that enhance its ability to cause disease, i.e., the ability to invade cells or to produce a toxin.

Virus: disease agent, smaller than bacteria, that is non-living and can only reproduce inside a host.

Vomiting: sudden and powerful emptying of stomach contents through the mouth.

Working Group on Civilian Biodefense: group consisting of twenty-five representatives from academia, medical research, military, government, public health, and emergency management that develop recommendations for procedures to be followed in case of use of various biological weapons.

***Yersinia*:** genus of bacteria that cause plague (*Y. pestis*) and gastroenteritis (*Y. enterocolitica*).

Zoonosis: a disease that can be transmitted from animals to humans.

REFERENCES AND RESOURCES

General

Books

Atlas, Ronald M. *Microbiology Fundamentals and Applications* (Macmillan, 1998, second edition).

Ayers, John C., J. Orwin Mundt, and William E. Sandine. *Microbiology of Foods* (W.H. Freeman and Company, 1980).

Brock, Thomas D. *Biology of Microorganisms* (Prentice-Hall, 1979, third edition).

Brown, Jack, Ph.D. *Don't Touch That Doorknob* (Byron Press Visual Publications, 2001).

Fox, Nichols. *Spoiled* (BasicBooks, 1997).

Garrett, Laurie. *The Coming Plague* (Farrar, Straus & Giroux and Harper Collins Canada Ltd., 1994).

Jay, James M. *Modern Food Microbiology* (Chapman and Hall, 1992, fourth edition).

Miller, Judith, Stephen Engelberg, and William Broad. *Germs: Biological Weapons and America's Secret War* (Simon and Schuster, 2001).

Satin, Morton. *Food Alert* (Checkmark Books, 1999).

Tierno, Philip M., Ph.D. *Protect Yourself Against Bioterrorism* (Pocket Books, 2002).

Winter, Ruth, M.S. *Poisons in Your Food* (Crown, 1991).

World Wide Web Sites

Eitzen, Edward; Julie Pavlin, et al. "Bacterial Agents" in *Medical Management of Biological Casualties Handbook*, U.S. Army Medical Research Institute of Infectious Diseases, Third Edition (1998):
www.nbc-med.org/SiteContent/MedRef /OnlineRef/FieldManual/medman/chap1.htm

Eitzen, Edward; Julie Pavlin, et al. "Biological Toxins" in *Medical Management of Biological Casualties Handbook*, U.S. Army Medical Research Institute of Infectious Diseases, Third Edition (1998):
www.nbcmed.org/SiteContent/MedRef /OnlineRef/FieldManuals/medman/chap3.htm

Centers for Disease Control and Prevention (CDC): **www.cdc.gov**

Christopher, George W. et al. "Biological Warfare, A Historical Perspective," *in Journal of the American Medical Association*, Vol. 278, No.5, pp. 412-417 (August 6, 1997):
http://jama.ama-assn.org/issues/v278n5/ffull/jsc7044.html

Environmental Protection Agency (EPA): **www.epa.gov**

Food and Drug Administration (FDA): **www.fda.gov**

Food and Drug Administration *The Bad Bug Book* at **www.cfsan.fda.gov/**

U.S. Army Medical Research Institute of Infectious Diseases (USAMRIID): **www.usamriid.army.mil/**
Virtual Naval Hospital: **www.vnh.org**

Resources by Chapter

Chapter One – *Bacillus anthracis*

Center for Civilian Biodefense Studies. *Anthrax* (2002):
www.hopkins-biodefense.org/pages/agents/agentanthrax.html

Inglesby, Thomas V. M.D., et al. "Anthrax as a Biological Weapon," in *Journal of the American Medical Association* Vol. 281, No. 18, pp 1735-1746 (May 12, 1999).

The Science of Terrorism, The Biology of Anthrax (2002):
www.jupiterscientific.org/sciinfo/sot.html

Sherman, Robert. Federation of American Scientists, Special Weapons Primer "Anthrax," (2002): **www.fas.org/nuke/intro/bw/agent.htm**

Smith, Michael, M.D., Web MD (AOL Health). *Anthrax: The Facts You Need* (November 2001):
http://aolsvc.health.webmd.aol.com/content/article/4058.323

Todar, Kenneth, Ph.D., "Todar's Online Textbook of Bacteriology," *Anthrax Toxin* (2002): **www.bact.wisc.edu/Bact330/lectureanthrax**

Chapter Two – *Salmonella*

Buckner, Rebecca. "U.S. Egg Safety Action Plan," in *Food Testing and Analysis*, (April/May 2000) pp. 8-10.

Cheng, Chorng-Ming et al. *PCR Assay for Rapid Detection of* Salmonella spp. *in Foods* (February 4, 2002): **www.cfsan.fda.gov/~frf/forum02/a088vpo.htm**

Holt, Peter S. *Public Meeting on* Salmonella enteritidis *Research,* September 8, 2000 (Transcript of proceedings): **www.foodsafety.gov/~dms/egg0900.html**

"Salmonellosis: no longer just a chicken and egg story," in *Canadian Medical Association Journal* Vol. 159, No. 63 (July 14, 1998): **www.cms.ca/cmaj/vol-159/issue-1/0063.htm**

Torok, T.J. et al. "A large community outbreak of Salmonellosis caused by intentional contamination of restaurant salad bars," in *Journal of the American Medical Association*, Vol. 278, No. 5, pp. 389-95 (Aug 6, 1997): **www.ncbi.nlm.gov/htbinpost /Entrez/query?uid=9244330&form=6&db=m&Dopt=b**

Weinberg, Winkler G., M.D. *About Salmonella* (2002): **www.about-salmonella.com/**

Wilson, Dick. *Salmonella vaccine on the way* (May 8, 1997): **www.cnn.com/TECH/9705/08/t_t/salmonella/**

Chapter Three – *Clostridium*

Arnon, Stephen S., M.D. "Botulinum Toxin as a Biological Weapon," in *Journal of the American Medical Association*, Vol. 285, No. 8, pp. 1059-1075 (February 28, 2001).

Botulinum Toxin, Center for Civilian Biodefense Studies (2000): **www.hopkins-biodefense.org/pages/agents/agentbotox.html**

Botulinum Toxins: **www.fas.org/nuke/intro/bw/agent.htm**

Facts about Botulism (October 2001): **www.bt.cdc.gov/documentsApp/FactSheet/Botulism/about.asp**

Nauman, Eileen, DHM. *Botulism as a Weapon of War* (October 16, 2001): **www.medicinegarden.com/Botulism_and_bioterrorism.html**

Sherman, Robert. Federation of American Scientists, Special Weapons Primer "Botulinum Toxin"(2002): **www.fas.org/nuke/intro/bw/agent.htm**

Wound Botulism. San Francisco Department of Health: **www.dph.sf.ca.us/healthinfo/woundbotulism.htm**

Chapter Four – Brucellosis and Q Fever

Brucellosis, Utah Department of Health, Bureau of Epidemiology (August 2001): **http://hlunix.hl.state.ut.us/els/epidemiology/epifacts/brucello.html**

"Brucellosis," in *Virtual Naval Hospital, Treatment of Biological Warfare Agent Casualties (Ch. 2: Bacterial Agents*), University of Iowa (July 17, 2000): **www.vnh.org/FM8284/Chapter2/2-11.html**

Chomel, Bruno B., M.D., et al., The American Veterinary Medical Association *Terrorism in the United States, Coxiella burnetii infection (Q fever)*(1995): **www.avma.org/press/terrorist_attack/zu_coxiella.asp**

"Q Fever," in *Virtual Naval Hospital, Treatment of Biological Warfare Agent Casualties (Ch. 2: Bacterial Agents)* University of Iowa (July 17, 2000): **www.vnh.org/FM8284/Chapter2/2-47.html**

Chapter Five – *Yersinia*

Inglesby, Thomas V., M.D., et al. "Plague as a Biological Weapon," in *Journal of the American Medical Association,* Vol. 283, No. 17, pp. 2281-2290 (May 3, 2000): **http://jama.ama-assn.org/issues/v283n17/labs/jst90013.html**

McGovern, Thomas W., M.D. and Arthur M. Friedlander, M.D. "Plague as a Biological Warfare Agent," in *Virtual Naval Hospital: Textbook of Military Medicine: Medical Aspects of Chemical and Biological Warfare:* **www.vnh.org/MedAspChemBioWar/chapters/chapter_23.htm**

Outbreak Control, Biological Agents Information Papers, United States Army Institute of Infectious Diseases (2002): **www.nbc-med.org/SiteContent/MedRef /OnlineRef/GovDocs/BioAgents.html**

Yersinia Pestis and The Black Death: **members.aol.com/omaruk/plague/**

Chapter Six – *Staphylococcus aureus*

Chesney, P.J., M.D., et al., *Epidemiologic Notes and Reports Toxic-Shock Syndrome – United States* (2002): **www.cdc.gov/mmwr/preview/mmwrhtml/00047818.htm**

Staphylococcal Enterotoxin B Disease, Biological Agents Information Papers, United States Army Institute of Infectious Diseases (2002): **www.nbc-med.org/SiteContent/MedRef/Gov/Docs/BioAgents.html**

Ulrich, Robert G., Ph.D., et al. "Staphylococcal Enterotoxin B and Related Pyrogenic Toxins," in *Virtual Naval Hospital, Textbook of Military Medicine: Medical Aspects of Chemical and Biological Warfare (Ch. 31:* Staphylococcal Enterotoxin B and Related Pyrogenic Toxins*)* (May 1997): **www.vnh.org/MedAspChemBioWar/chapters/chapter_31.htm**

Chapter Seven – Cryptosporidiosis, Cholera, and Tularemia

Dennis, David T., M.D., et al. "Tularemia as a Biological Weapon," in *Journal of the American Medical Association* Vol. 285, No. 21, pp. 2763-2773 (June 6, 2001): **http://jama.ama-assn.org/issues/v285n21/ffull/jst10001.html**

Evans, Martin E., M.D. and Arthur M. Friedlander, M.D. "Tularemia," in *Virtual Naval Hospital, Textbook of Military Medicine: Medical Aspects of Chemical and Biological Warfare (Ch.* 24 – Tularemia) (May 1997): **www.vnh.org/MedAspChemBioWar/chapters/chapter_24.htm**

Tularemia, Biological Agents Information Papers, United States Army Institute of Infectious Diseases (2002): **www.nbc-med.org/SiteContent/MedRef/Gov/Docs/BioAgents.html**

Chapter Eight – *E. coli* O157:H7 and *Shigella*
Kolavic, Shellie A. ,et al. "An Outbreak of *Shigella dysentariae* Type 2 among Laboratory Workers Due to Intentional Food Contamination," in *Journal of the American Medical Association* Vol. 278, No.5, pp. 396-398 (August 6, 1997): **http://jama.ama-assn.org/issues/v278n5/ffull/joc71418.html**

Mermelstein, Neil H. "Controlling *E. coli* O157:H7 in Meat," in *Food Technology* (April 1993) pp. 90-91.

Muller, Rainer. "What is *E. coli* O157:H7?"in *Eric's Echo* (1998): **www.ericsecho.org/whatisec.htm**

Chapter Nine – Food-borne Viruses and Other Viral Agents
Borio, Luciana M.D., et al. "Hemorrhagic Fever Viruses as Biological Weapons," in *Journal of the American Medical Association*, Vol. 287, No. 18, pp. 2391-2405 (May 8, 2002): **http://jama.ama-assn.org/issues/v287n18/ffull/jst20006.html**

Henderson, Donald A., M.D., et al. "Smallpox as a Biological Weapon," in *Journal of the American Medical Association*, Vol. 281, No. 22, pp. 2127-2137 (June 9, 1999): **http://jama.ama-assn.org/issues/v281n22/ffull/ist90000.html**

McClain, David J., M.D. "Smallpox," in *Virtual Naval Hospital, Textbook of Military Medicine: Medical Aspects of Chemical and Biological Warfare (Ch. 27,* Smallpox) (May 1997): **www.vnh.org/MedAspChemBioWar/chapter/chapter_27.htm**

Smallpox Vaccination Decision Near. Associated Press (Sunday, July 7, 2002): **www.nlm.nih.gov/medlineplus/news/fullstory_8411.html**

Smith, Jonathan F., Ph.D., et al. "Viral Encephalitides," *in Virtual Naval Hospital, Textbook of Military Medicine: Medical Aspects of Chemical and Biological Warfare (Ch. 28)*: **www.vnh.org/MedAspChemBioWar/chapter/chapter_28.htm**

Variola (Smallpox) Biological Agents Information Papers, United States Army Institute of Infectious Diseases (2002): **www.nbc-med.org/SiteContent/MedRed/Gov/Docs/BioAgents.html**

Eitzen, Edward; Julie Pavlin, et al. *Medical Management of Biological Casualties Handbook,* U.S. Army Medical Research Institute of Infectious Diseases, Third Edition (1998). **www.nbcmed.org/SiteContent/MedRef /OnlineRef/FieldManuals/medman/chap2.htm**

Chapter Ten – Natural Toxins

Ricin Intoxication, Biological Agents Information Papers, United States Army Institute of Infectious Diseases (2002):
www.nbc-med.org/SiteContent/MedRef /OnlineRef/GovDocs/BioAgents.html

Wannemacher, Robert W., Ph.D. and Stanley L. Wiener, M.D. "Tricothecene Mycotoxins," in *Virtual Naval Hospital, Textbook of Military Medicine: Medical Aspects of Chemical and Biological Warfare (Ch. 34* Tricothecene Mycotoxins) (May 1997):
www.vnh.org/MedAspChemBioWar/chapters/chapter_34.htm

Chapter Eleven – Chemicals and Pesticides

Agency for Toxic Substances and Disease Registry, *Mustard Gas* (September, 1995): **www.atsdr.cdc.gov/tfacts49.html**

National Institute of Environmental Health Services, *Pesticides* (March 13, 2002): **www.niehs.nih.gov/external/faq/pest.htm**

Water Resource Characterization DSS – Heavy Metals:
www.h2osparc.wq.ncsu.edu/info/hmetals.htm

Chapter Twelve – BSE and CJD

Chronic Wasting Disease, Minnesota Department of Natural Resources, Division of Wildlife (March 2002): **www.dnr.state.mn.us/hot_topics/030102.html**

Facts About Chronic Wasting Disease, Missouri Department of Agriculture, Division of Animal Health (June 14, 2002):
www.conservation.state.mo.us/hunt/deer/cwd.htm

First live test for chronic wasting disease succeeds, Environmental News Network (December 12, 2001):
www.env.com/news/enn-stories/2001/12122001/s_45849.asp

Lyman, Howard F and Glen Merzer. *Mad Cowboy* (Simon and Schuster, 1998).

Chapter Thirteen – Agri-Terrorism (H3)

Agricultural Terrorism, DomesticWMD (Weapons of Mass Destruction) Terrorism: Assessing the Threat (March 2000):
www.infowar.com/class_3/00/class3_tp-terr.shtml

Ban, Jonathan. *Agricultural Biological Warfare: An Overview* (June 2000):
www.mipt.org/agterror-rpt.html

Bioterrorism, National Animal Health Emergency Management System:
www.usaha.org/NAHEMS/bioterr.html

Cameron, Gavin, Jason Pate, and Kathleen M. Vogel. "Planting Fear, How real is the threat of agricultural terrorism?" in *Bulletin of the Atomic Scientists* (2001) Vol. 57, No. 5, pp. 38-44:
www.thebulletin.org/issues/2001/so01/so01vogel.html

Chalk, Peter. "The Threat Beyond 2000," in conference proceedings from *Bioterrorism: Homeland Defense: The Next Steps* (Threat Panel) (February 2000):
www.mipt.org/agterror-rpt.html

Foot-and-Mouth-Disease, Research Articles for the Pork Producer Swine Veterinarian and Pig Owner (2001):
www.thepigsite.com/FeaturedArticle
   /Default.asp?AREA=FeaturedArticle&Display=305

Knebusch, Kurt. *Bioterrorism May Be Threat to U.S. Agriculture, Expert Says* (August 2001) (Ohio State University website):
www.osu.edu/units/research/archive/croppat1.htm

Q&A: Foot-and-Mouth Disease, CNN.com/World (February 26, 2001):
www.cnn.com/2001?WORLD/europe/UK/02/21/foot.mouth/

Wilde, Matthew. "After Sept. 11, the safety of the U.S. food supply can no longer be taken for granted," *Courier* (October 7, 2001):
www.wcfcourier.com/Metro2001/011007after.html

Chapter Fourteen – Working Together

Backgrounder – Food Safety and Food-borne Illness, International Food Information Council (August 1998):
www.ificinfo.health.org/backgrnd/bkgr10.htm

Biological, chemical weapons: Arm yourself with information (1998-2002):
www.mayoclinic.com/invoke.cfm?id=MH00027

Food Safety and Inspection Service. *Biosecurity and the Food Supply* (June 2002): www.fsis.usda.gov/oa/background/biosecurity.htm

Bioterrorism, Georgia Department of Public Health (2000):
www.ph.dhr.state.ga.us/programs/emerprep/biogeneral.shtml/

Bush signs bioterror law, pushes homeland security (CNN.com, June 12, 2002):
http://www.cnn.com/2002/ALLPOLITICS/06/12/bush.terror/

Cramer, Michael. *Bioterrorism: The Next Food Safety Threat,* (October 2001):
www.ces.ncsu.edu/depts/foodsci/agentinfo/hot/bioterrorism.html

Food Safety and America's Future, National Food Processors Association meeting, (November 27, 2001): www.hhs.gov/news/speech/2001/011127.html

Food Supply Vulnerable to Attack, (CBS News, October 11, 2001):
www.cbsnews.com/now/story/0.1597.314454-412.00.shtml

Harrison, Alisa. *USDA Continues to Step Up Homeland Security Efforts* (April 30, 2002): **www.usda.gov/news/releases/2002/04/0176.htm**

HHS Initiative Prepares for Possible Bioterrorism Threat (August 6, 2001): **www.hhs.gov/news/press/2001pres/01fsbioterrorism.html**

How Safe is Our Food? (October 2001): **www.nutrition.about.com/library/weekly/aa102601a.htm**

Meserve, Jeanne. *FDA to propose antiterrorist food safety guidelines,* CNN (January 8, 2002): **www.asia.cnn.com/2002/us/01/08/rec.food.safety.terrorists**

NFPA Says FDA Food Security Guidance Strengthens Barriers to Food Security Threats, National Food Processors Association News Release (January 8, 2002): **www.nfpa-food.org/newsreleases/newsrelease010802.htm**

Testimony of the Honorable Ann M. Veneman, Secretary of Agriculture Before the United States Senate Committee on Agriculture, Nutrition and Forestry (July 17, 2002): **www.usda.gov/news/special/ctc31.htm**

USDA Releases $4.3 Million to States for Strengthening Agriculture Homeland Security Protections (May 30, 2002): **www.usda.gov/news/releases/2002/05/0213.htm**

INDEX

This index contains references only to textual materials--no graphics or tables are included. At the beginning of each chapter is a line drawing of the organism discussed in that chapter. Likewise at the end of each chapter is a table noting the organism, the diseases it causes, symptoms, mode of transmission, prevention and treatment measures, and the organism's potential as a bioweapon in food or water. Scientific names are included in the index and where possible refer the reader to the common name of the organism or illness which is where the reader will find page references to the text.

BOOK ORDER FORM

❑ **YES!** I would like to order *FOOD FIGHT: The Battle to Protect our Food and Water Against Terrorism*

Name _____

Organization _____

Address ❑ Work ❑ Home _____

City _____ State_____ Zip_____

Home Phone _____ Work _____

Fax_____ Website_____

Email _____

Quantity ordering _____ x $19^{95} = $_____

WI sales tax: add 5.5% = $_____

Shipping and Handling per book $4^{00} = $_____

ISBN: 0-9722099-4-8   **TOTAL** $_____

All orders pre-paid please!

❑ By credit card: ❑ VISA ❑ Mastercard

Card #:_____ - _____ - _____ - _____

Expiration Date: _____ Signature_____

❑ By check: Payable to **Goblin Fern Press**
222 N. Midvale Boulevard
Suite 24
Madison, WI 53705

Orders may be submitted by mail, fax or
through our secure website, **www.goblinfernpress.com**.

Tel: 608.218.1646 • Fax: 608.218.1647 • Email: info@goblinfernpress.com

BOOK ORDER FORM

❑ **YES!** I would like to order *FOOD FIGHT: The Battle to Protect our Food and Water Against Terrorism*

Name _____

Organization _____

Address ❑ Work ❑ Home _____

City _____ State _____ Zip _____

Home Phone _____ Work _____

Fax _____ Website _____

Email _____

Quantity ordering _____ x $19⁹⁵ = $_____

WI sales tax: add 5.5% = $_____

Shipping and Handling per book $4⁰⁰ = $_____

ISBN: 0-9722099-4-8	**TOTAL** $_____

All orders pre-paid please!

❑ By credit card: ❑ VISA ❑ Mastercard

Card #:_____ - _____ - _____ - _____

Expiration Date: _____ Signature_____

❑ By check: Payable to **Goblin Fern Press**
222 N. Midvale Boulevard
Suite 24
Madison, WI 53705

Orders may be submitted by mail, fax or
through our secure website, **www.goblinfernpress.com**.

Tel: 608.218.1646 • Fax: 608.218.1647 • Email: info@goblinfernpress.com